American Science Policy
Since World War II

About Brookings

The Brookings Institution is a private nonprofit organization devoted to research, education, and publication on important issues of domestic and foreign policy. Its principal purpose is to bring knowledge to bear on the current and emerging policy problems facing the American people.

A board of trustees is responsible for general supervision of the Institution and safeguarding of its independence. The president is the chief administrative officer and bears final responsibility for the decision to publish a manuscript as a Brookings book. In reaching this judgment, the president is advised by the director of the appropriate Brookings research program and a panel of expert readers who report in confidence on the quality of the work. Publication of a work signifies that it is deemed a competent treatment worthy of public consideration but does not imply endorsement of conclusions or recommendations. The Institution itself does not take positions on policy issues.

AMERICAN SCIENCE POLICY SINCE WORLD WAR II

Bruce L. R. Smith

The Brookings Institution
Washington, D.C.

Copyright © 1990 by
THE BROOKINGS INSTITUTION
1775 Massachusetts Avenue, N.W., Washington, D.C. 20036

Library of Congress Cataloging-in-Publication Data

Smith, Bruce L. R.
    American science policy since World War II / Bruce L. R. Smith.
    p. cm.
    ISBN 0-8157-7998-4 (alk. paper) ISBN 0-8157-7997-6
(pbk.: alk. paper)
    1. Science and state—United States. 2. Technology and state—
United States. I. Title.
Q127.U6S613   1990
338.97306—dc20   90-31421
                                                CIP

9 8 7 6 5 4 3

The paper used in this publication meets the minimum requirements
of the American National Standard for Information Sciences—
Permanence of Paper for Printed Library Materials, ANSI Z39.48-
1984.

Set in Linotron Sabon with Helvetica Display
Composition by Graphic Composition, Inc.
  Athens, Georgia
Printing by R. R. Donnelley and Sons, Co.
  Harrisonburg, Virginia
Book design by Ken Sabol

Preface

JUST AFTER THE CLOSE of World War II, America's political and scientific leaders reached an informal consensus on how science could best serve the nation and how government might best support science. The consensus lasted a generation before it broke under the pressures created by the Vietnam War. Since then the nation has struggled to reestablish shared beliefs about the means and goals of science policy. This book is an effort to make sense of that experience, to identify the patterns in postwar science affairs, and to show that what might otherwise seem a miscellany of separate episodes actually constituted a continuing debate of national importance that was closely linked to broad political and economic trends. The hope is that a better understanding of where we are and how we got there might cast modest light on future policy directions.

I have discussed science policy and the book's major themes with hundreds of friends and colleagues over many years. There are so many who have made important contributions that it seems almost impossible to single them out. Yet some contributions have truly been crucial. Harvey Brooks of Harvard University read the entire manuscript closely, saved me from many errors, and made connections where my imagination failed. Similarly, Robert Rosenzweig of the Association of American Universities read the manuscript with a keen eye and offered many valuable suggestions.

The following Brookings colleagues read the entire manuscript or parts of it and made comments and suggestions or otherwise provided useful assistance: Bruce MacLaury, Thomas Mann, Edward Denison, James Carroll, Lawrence Korb, Thomas McNaugher, Pietro Nivola, James Reichley, Walter Beach, Kenneth Flamm, and Robert Faherty. Donald Frey of Northwestern University made especially valuable contributions to my understanding of industrial research and its relation to corporate strategy. John Dealy of Georgetown University taught me more than I can easily acknowledge about business and about defense research. Rodney Nichols helped me understand the biomedical research

v

community and mission-oriented research and suggested revisions that greatly improved the final chapter. Carlos Kruytbosch of the National Science Foundation and Roger Porter of the National Institutes of Health cheerfully responded to numerous requests for data and contributed generously of their time and knowledge of science affairs. Edward Shils made me keep trying.

James Schneider edited the manuscript with great skill. Susan McGrath and her colleagues in the Brookings library tracked down numerous obscure sources and references. Susan Popkin, John Henry Fullen, Adam Weingard, Anthony Maybury-Lewis, Diana Turacek, and Linda Miller provided research assistance at various stages. Susan Williams and Marjorie Crow typed and retyped drafts. Mary Isaacson proofread the manuscript, Susan Woollen prepared it for typesetting, and Margaret Lynch compiled the index. I thank them all while absolving them of responsibility for any remaining errors or shortcomings.

Of course, the views that are expressed in this book are my own and should not be ascribed to the trustees, officers, or other staff members of the Brookings Institution.

BRUCE L. R. SMITH

March 1990
Washington, D.C.

Contents

Tables

Figures

To Elise

1 ||| Science Policy in the American Context

A DISTINCTIVE FEATURE of American government since World War II has been the emergence of "science policy" as a focus of thought and action. Before the war, numerous obstacles, institutional and traditional, worked against a significant role for science and technology in public affairs. First, the Constitution has always restricted the powers of the federal government to minimum necessary functions, only two of which have a direct bearing on the mobilization, regulation, or coordination of scientific resources. Article I, section 8 empowers Congress "to promote the progress of science . . . by securing to authors and inventors the exclusive rights to their respective writings and discoveries," but for a limited time only. The same article also entitles Congress to prescribe "uniform weights and measures," a power that eventuated in the establishment of the National Bureau of Standards in 1901. Beyond these functions, all matters relating to education, interstate commerce, internal improvements, and culture are left to the states (and, of course, many powers were thought by the framers to be beyond the ken of government at all).

Government activities, to be sure, encroached on these boundaries during the nineteenth century. "Necessary and proper" measures to carry out enumerated powers enlarged the federal government's scope. As the nation expanded across the continent, no static formula could contain the living reality of the governing process, and exercising its powers naturally involved the government in matters of science and technology. For instance, the efforts of the Army Ordnance Department in the 1840s to develop better weapons led the military to promote standardized manufacture and the use of interchangeable parts. This example illustrates the difficulty—still a source of conceptual confusion—of separating science and technology from the wider social ends to which they are but means. The army was not interested in technological innovation for its own sake but recognized that it could not have certain improvements to weapons

1

and wagons without expanding the nation's scientific and technical knowledge base.

Almost from the beginning of the Republic, the federal government surveyed and mapped the land, collected customs, watched over public health, and pursued other technical activities in specialized agencies staffed with professionals from relevant scientific fields. Science and technology were familiar enough to the Founders and succeeding generations of politicians and government officials, and were regarded as useful for carrying out accepted tasks more efficiently. Scientific activities were, however, usually not seen as matters of central importance to the great public issues of the day. In only a few instances, most notably perhaps in supporting the growth of American agriculture, did a conscious strategy link research systematically to achieving broad policy goals. Beginning with the Civil War, research efforts also became an integral part of the military's mission (though technological change tended to return to a leisurely pace following the war). But such cases of science and technology as the centerpieces of government action or the motivating forces behind new programs or agencies were rare before World War II.

Even as the assumptions underlying government activity began to change and the elements of a modern welfare state began to take shape, scientific research remained a peripheral activity. Care for the needy, shelter, universal education and literacy, health, and other aspects of social policy became objects of government action, but direct government support for science in the universities, or the direct linkage of research activities to the fulfillment of broad social goals, came more slowly as an element of the highly mobile societies of our own time.

Scientific inquiry, like the highest forms of the arts, was nurtured by private patronage. Indeed, many of its practitioners, though now often considered political liberals, were radical critics of government. Joseph Priestley, the great English chemist and dissenter, was representative of this radical Baconianism. For Priestley, the state was the enemy of scientific progress, economic growth, and religious freedom, and was fit only to be swept away in the full flowering of the human spirit that would accompany the scientific revolution.[1] Such views influenced the bias of the American Constitution toward limited government. It is perhaps only logical, then, that for much of America's history the sciences developed without direct government patronage. Most academic scientists remained suspicious of federal support for science well into the 1930s.

The Second World War, however, brought new assumptions, institutions, and policies. This book tells how the nation responded to the altered relationship between government and science after the war and especially how changing circumstances have continued to test the arrangements. The ensuing chapters emphasize challenges to the American consensus on the appropriate roles of science, technology, and government. What was the nature of the bargain struck when it became obvious that the nation would need to respond somehow to the new importance of science in public affairs? What challenges and stresses arose as the postwar system grew in size and complexity, as the world economy became more competitive, dynamic, and interdependent, and as the framework of national security was transformed? What remains valid in the early conceptions of how to relate science effectively to government, and what should be modified or discarded?

The first phase of the relationship lasted from about 1950, when the postwar system was put in place, until the system became subject to increasing stresses and began to undergo a transformation about 1966. A second phase ensued until the middle of the Carter administration, when the system entered a third, and current, phase.[2]

In the first phase, the nation celebrated the impact of science in terms resembling Priestley's view of its liberating influence on human affairs. The need for policy amounted mainly to devising the means to support basic research, to link applied research effectively to national priorities, to coordinate the jurisdictional issues that inevitably arise in the wake of scientific and technological advance, and to regulate a few obviously dangerous technologies while, of course, promoting beneficial developments. Defense, space, and atomic energy issues dominated this phase.

Phase two saw a questioning of many assumptions of the first phase. A darker vision replaced the innocence and optimism. Regulating dangerous side effects of technology became an urgent challenge. Such social priorities as preserving the environment and protecting consumers became important objectives. Critics, curiously, assumed that technology could be easily redirected to serve more humane ends, and they sometimes presented their own versions of simple technological solutions to complex problems—such as the mandates to use "best available technology" that are common in legislation to control pollution. The progression from basic research to practical application also proved less automatic than previously supposed. More policy attention was required to move a discovery to eventual commercialization. Government support for research slowed. The free flow of ideas and open scientific communication—previously a self-evident and undisputed good—aroused some uneasiness in the light of changing national security needs and the lessened competitiveness of American firms.

Fears about the beneficence of scientific inquiry and technological innovation receded, however, and the nation recovered its faith in science and technology at the end of the 1970s. But there has been no tidy synthesis reconciling the divergent tendencies of the earlier periods. The current phase of government-science relations in America embodies many of the conflicts of the past and has left unresolved some of the most critical issues.

Some Boundaries

The periods into which I divide postwar relations between government and science reflect, of course, my sense of what is significant in a subject matter of great scope and unclear boundaries. I hope that the division into three periods avoids arbitrariness, but history is always resistant to pat formulas, and points of demarcation do not spring conveniently from the welter of events. Science affairs constitute a subject matter more resistant than most to conceptual clarity. In any event this scheme makes a convenient peg on which to hang a discussion of critical issues.

Despite a vast outpouring of literature, the core of "science policy" or "science affairs" has proved elusive. One typology pronounced the subject as ranging from "policy for science" to "science in policy."[3] This was a helpful distinction, but the two perspectives tend to become either too narrow or too broad. In one, the observer could get caught up in the minutiae and mechanics of research support and succumb to triviality. In the other the subject could dissolve so that science becomes merely one of the myriad influences playing a small part in creating policies. Wallace S. Sayre, indeed, depicted scientists as constituting merely another group that sought to advance its own interests or conception of the public good on the basis of claims to special expertise. The role of scientists in determining policy should be demystified, and they should recognize that they were "in the battle, and not above it."[4] Jürgen Schmandt and James Everett Katz, on the other hand, saw the emergence of a new "scientific state" different in character from the "administrative state" of the industrial revolution.[5]

Don K. Price tried to reconcile the two points of view and to account for the influence of science by references to American traditions and values, yet show the revolutionary impact on the nation's governing institutions.[6] He saw scientists as interacting with professionals, administrators, and politicians as scientific developments move from the early stages dominated by specialists to the wider political arena in which narrow expertise blends with broad judgments on values, constitutional constraints, and political norms. This spectrum of activity is reflected in the difference between universities and other research institutions and the government. The difference is also found within the formal government hierarchy in the contrast between the so-called mission agencies and the staff units surrounding the president. The agencies or line departments such as Defense, Transportation, Agriculture, and Health and Human Services pursue technological and research programs to advance public goals, while the staff units of the president and Congress fit those programs into the framework of presidential and congressional priorities. Science and technology help give our governing system its striking dynamism, but they operate within a web of values, customs, and norms that give the whole process stability and orderliness.

There have been other broad interpretations of American science policy by historians, sociologists, political scientists, lawyers, economists, and physical scientists, but for the most part, as the study of science affairs developed and as research mushroomed, the field became a cluster of loosely related subspecialties.[7] There is, of course, nothing new or remarkable in this pattern so often found as scientific inquiry explodes in new directions. "The Endless Frontier" is the apt phrase that Vannevar Bush used to characterize the open-ended process of scientific exploration that would replace the western frontier in the American consciousness.[8]

Science affairs, however, never had a disciplinary core to begin with and constituted perhaps more a profession than a field. That is to say, those who were facing practical problems involving science and technology drew upon whatever experience or line of inquiry seemed to shed light. The field thus became highly eclectic, with contributions from law, economics, management, the physical sciences and engineering, as well as interdisciplinary inquiry of all kinds. Yet there are times when study benefits from an effort to relate the parts to the whole, when the specialized tributaries are viewed from the perspective of their contribution to the central system. Inquiry can be occasionally energized by a restatement of core concepts. A modest effort in that direction, the aim here is to clarify the main issues that have concerned observers of science policy in the postwar period and thus to remind us of common reference points for the many lines of inquiry.

Science policy is concerned with the *promotion* of scientific discovery and technlogical innovation, the *regulation* of potentially harmful side effects resulting from their application, and the *coordination* of policies and programs within the government and in society to achieve the appropriate balance between nurture and restraint. At any one time, government policy may emphasize one or another objective, but typically promotion, regulation, and coordination are present in some combination in every government science and technology program. Even a regulatory agency finds that it cannot sensibly carry out its duties without performing research either in its own laboratories or in the outside institutions it supports through grants or contracts (or, typically, both). Rare is the agency that does not seek to employ science and technology to fulfill its responsibilities or to develop new missions. So, too, the business firm uses its research department to invent new products, and the university laboratory seeks to advance itself through discovering new techniques or developing new fields of inquiry.

More particularly, science policy is made up of the following elements:

—*Basic research.* The nation's science policy is a mix of policies, programs, and institutional arrangements that has evolved to create and sustain basic research. How satisfactorily has the system worked to promote scientific and technological advance to achieve broad social goals?

—*Applied research.* What system has evolved to perform the research deemed necessary by government agencies? What assumptions have guided applied research efforts and how successfully has this part of the research system performed? Are there logical and mutually reinforcing links between the institutions and programs involved with basic research and those with applied research?

—*Commercialization.* How does scientific and technological research become translated into new products and services? Commercialization involves a range of activities, some highly technical, as a scientific idea becomes embodied in production and distribution, enhancing economic growth, creating jobs, and keeping American industry competitive. What policies, if any, should the government undertake to remedy shortcomings, remove obstacles, or otherwise ensure that science and technology contribute to a strong economy? Has the government's role been significantly modified by the increasing importance of foreign trade and of international capital and technology flows?

—*Regulatory behavior.* What assumptions have shaped government agencies' use of science and technology to assess risks to health, safety, or environmental quality? Regulatory policy, of course, extends considerably beyond science policy, but depends on the use of science in setting standards, determining priorities for regulation, and measuring pollutant concentration. How do scientific aspects of regulatory policies relate to policies that seek to promote technical advance and the commercial exploitation of technology? Can a reasonable balance be struck between promotion and regulation when these objectives conflict?

—*International issues.* Scientific knowledge influences public policy on such matters as arms control, the choice of weapons systems, the development of institutions for managing satellite communications, and intellectual property rights in the transfer of technology across national borders. Policies toward scientific inquiry include the support of science abroad and cooperation in large scientific undertakings. What special complexities for science and technology policy result from the international political environment and from national security imperatives? Can the nation pursue a guns and butter strategy successfully with respect to its scientific resources—that is, can it accommodate military and civilian technology needs, strategic and commercial objectives, large research programs and the requirements of a healthy body of small projects?

Science and Values

Science policy is not and cannot be the subject of value-free inquiry. To answer the practical questions, of course, one must understand the way technology diffuses through the economy and society, the connections between basic and applied research, and the behavior of complex organizations. The intellectual challenge and excitement of the study of science policy, with its intimate connections across the broad domain of learning and its tributaries of specialized knowledge, cannot be denied. Nonetheless such study is at bottom an applied field, a part of the policy sciences concerned with decisionmaking in federal and state governments, businesses, and universities.[9] What brings its practitioners together is the desire to understand how science and technology affect society, how they can be made to serve human ends more effectively, and how their nurture can best be accomplished with such broader goals in mind.

It follows that no aspect of the broad field is ever wholly removed from the interest or the attention of citizens. It may seem strange to have nonscientists serve on a National Institutes of Health study section or consensus panel, to have city council members judging whether *E. coli* studied in a university laboratory constitutes a menace to the community, to see state legislators ruling on Darwinism versus creationism, or to observe cities declaring themselves nuclear free zones. There are certainly instances of spectacular nonsense resulting from popular participation in discussing technical issues. One state legislature is alleged to have sought to establish a streamlined value for the mathematical expression *pi*. The House of Representatives once tried to subject each of approximately 14,000 National Science Foundation grants to a potential legislative veto. Legislative vetos have been declared unconstitutional, but the reality of citizen participation and the involvement of elected representatives remains.

Such episodes reflect an underlying truth about American politics and the American constitutional system: citizens, or those who speak on their behalf, are entitled to intervene in any aspect of policy that they choose, and they frequently do so, even on matters that experts regard as their special province. If this means that policymaking is disorderly and subject to demogogic manipulation, the system holds these to be lesser evils than to allow only so-called experts to define policy and thus monopolize the details of administration. The British tradition of a civil service that dominates administrative matters except for clearly defined instances of ministerial discretion is rejected under America's unwritten constitution, which eschews any establishment, civil or religious.

The most arcane area of scientific judgment, even the allocation of resources within well-defined disciplines, may become subject to scrutiny. Politicians may wish to assess the likelihood of practical applications from the investment of public support, to require researchers to observe conflict-of-interest rules, to judge whether research funds are equitably distributed geographically, or merely to inform themselves about the progress of a discipline. There is no entitlement to continued support from public funds for any scientific discipline, institution, program, or for science and technology as a whole. The scientific enterprise must justify itself by effective use of resources and contribution to society. The crusades against scientific fraud that sometimes reach even the most distinguished scientists and prestigious institutions illustrate that science is not exempt from the rough and tumble of politics. But it is understandable that people are angered by fraud in science, and it is crucial that science keep its house in order. Scientific fraud is perhaps the gravest threat to science because the automony from political interference that it has enjoyed for forty years is based on widespread trust in the integrity of the system and in the willingness of scientists to discipline themselves.

Some scientists have come to believe, after the rapid growth of R&D budgets following World War II, that their fields or institutions are entitled to continued public support at least at some level. Society has been generous in supporting research, but that support is not an entitlement program mandated by statute. Research activities will compete against each other and against other classes of activity in the budgets of companies, states, and the federal government. And competition has grown more severe, which helps account for the paradox that scientific investigators have felt squeezed even as policymakers have sometimes grown weary of the seemingly inexorable rise of budget outlays for science and technology.

For the near future, a large budget deficit, balance of trade and payments problems, and the need for spending restraint make the outlook dim for new government programs. Scientists will do well to make their case for public support in terms of the social values they serve, not by the pretense that they are members of a special guild immune from outside scrutiny.

Science and Democracy

Science policy has become more accessible to criticism by the public and more widely debated during the postwar period. For scientists to wrap themselves in the mystique of their craft, to resist immersing themselves in the muddy issues of policy, or to block public involvement would fly in the face of the logic of their situation. Perhaps more than ever they need public support for the enterprise of science and public understanding of what it can and cannot contribute to national objectives. If scientists are sometimes uncomfortable venturing beyond their immediate expertise, they may take comfort in challenging the hucksterism, shrillness, and plain lack of information that often distort public debate of complex technical and related policy issues.

Taking a responsible part in debates on policy as well as educating people in the workings of science will not necessarily come easily. It is an honored tradition for scientists to eschew loosely formulated issues lacking a sharp focus and to entrust authority to those with the most specialized knowledge of a subject. In the pure vision of science there is scarcely a role for general authority. The enterprise goes forward more or less autonomously as problems are broken down into more specific ones and experts lead in deciding what to do next.

Michael Polanyi once evoked the vision of a Republic of Science where the informal interactions among scientists result in more deeply rational choices than could any effort to guide from the outside, for "any attempt at guiding scientific research toward a purpose other than its own is an attempt to deflect it from the advancement of science" and is bound to fail because science "can advance only by essentially unpredictable steps. . . . You can kill or mutilate the advance of science, you cannot shape it." [10]

Polanyi's vision resembles Priestley's and Bacon's views of how science and technology fit into society. Just as authority is parceled out on the basis of individual expertise among self-governing colleagues, so too in the social division of labor the role of central authority is sharply limited. The most competent people, they would argue, should perform tasks according to their specializations. Government (or power) is corrosive, unnecessary, reflective of the dead hand of the past. It obstructs the technical progress that could liberate mankind. This is a vision of society far more radical than that of nineteenth century Marxism. Its impact on American constitutionalism is unmistakable. James Madison's "new science of politics" is based on the idea that rationality and empirical inquiry should substitute for religion and revealed truth as the basis for political action. The development of scientific ideas by the various departments of government and by outside groups was to be the source of energy for the political system.

The loose pluralism of the institutional system, once the nation had shed the theocracies of the colonial era, inevitably impelled people and institutions to persuade, conciliate, and cooperate with each other. As the system evolved, powerholders shared the power-averse attitudes of their fellow citizens. [11] The civil service, instead of becoming an establishment in the tradition of Whitehall or the Continental bureaucracies and providing unified control over the apparatus of government, embodied specialization in its top ranks and resisted direction from the center. Politically neutral expertise was the route to advancement in a system in which the parts warred against the whole. The career bureaucrat was as irritated by central government authority as the academic scientist by the campus administration or the industrial R&D staff by the bean counters in the front office.

Polanyi and the purists are correct in insisting on the autonomy of science as a community of specialized professionals and in asserting their right to make certain kinds of decisions without outside interference. Judgments on the merit of scientists, the value of research proposals within a well-established field of basic research, and tactical questions regarding how projects are to be carried out are best left to scientists themselves.

But this isolation is inappropriate when one leaves the realm of pure science and enters that of technological applications. And even within pure science, strategic and tactical choices merge.[12] Choices on support for new fields of inquiry, support for overlapping and interdisciplinary studies, and support for different categories of activity do not have scientific peer reviewers in any conventional sense to decide precisely framed questions. Broader and more inclusive judgments than the scientific merit of an individual or of research proposals within a subdiscipline must be made regularly. Some measure of planning is inescapable within fields, between fields, among classes of scientific investigators, and among institutions.[13] Equilibrium could exist only in a static world. Of course, the dynamism of science virtually guarantees the emergence of new fields, more sophisticated instruments, and new opportunities for applying technology. These developments engage the attention of scientists in other fields, administrators, budget officials, chief executives, legislators, and citizens.

The scale of scientific activity has also changed dramatically as more and more fields have moved toward "big science" (highly capitalized efforts with steep start-up and maintenance costs). This change has meant that judgments broader than the merits of one investigator must be carefully weighed. As Donald Kennedy, president of Stanford University, observed,

In the 1960's, the main problem for department heads in making appointments had to do with billets; that is, with persuading the dean and the provost that the growth potential of our discipline justified the retention of an additional faculty person. A decade later . . . the main problem became whether we could commit the equipment and space renovations necessary for the new appointee to do the work. The capital cost of the equipment and special facilities, in short, had become larger than the capital value of the endowment necessary to yield the faculty member's salary.[14]

Clearly, the tactical and strategic dimensions of scientific choice are brought closer together by such developments. Critical decisions increasingly draw upon a wide circle of experts and administrators. It is intriguing to speculate about the impact of changes in the scientific climate on the motives that drive it. As a social system, science consists of individuals impelled by a reward system based on priorities of discovery.[15] How may this classic system of "little science" be affected if resource decisions and access to major facilities create inequities? The circumstances of modern science belie the image of an autonomous, pure Republic of Science where people automatically defer to those most knowledgeable in each subject and where perfect harmony prevails and power and authority are unknown.

But for its continued appeal in American thought, the radical vision of science as the handmaiden of social harmony would require little comment. Although they know better, Americans seem perennially tempted by the allure of a society apparently free of hierarchy, whether in the church, family, or world order, whether in a commune trying to embody the spirit of early Christians, a metropolitan government with power dispersed among single-purpose functional authorities, a political party run by the "people," policy made by referendum, control of local schools by the community, or the self-governing factory where each worker realizes his or her full potential.

The impulse for society without government and for government without coercion is rooted in the only absolute of America's pluralist doctrine: the idea of individual liberty. The maximum freedom consistent with political order is tied to the corollary that power must be limited. To be limited means that it must be shared, narrow in scope, dispersed institutionally, and exercised lightly—and with multiple checks. These abiding values, however, are not served by the pretense that power is absent from human affairs, that it can be willed away, or that it is more bearable when accompanied by hypocrisy and self-deception. The more practical side of the American tradition is much to be preferred to the utopian impulse or reformist crusade.

Perhaps the most common source of confusion is the assumption that science and democracy are, or should strive to be, alike. But the processes and the norms by which science operates are profoundly unlike political ones, even in democratic nations. Science and democracy are something like marriage partners who get along best when they respect each other's differences.[16] The elitism of science is compatible with democracy if understood in this context.

Science and the Public Policy Agenda

Scientists will always be the ones most interested in scientific work and in the circumstances affecting the progress of science. Although they have often become intensely interested in policy (just as nonscientists have interested themselves in science), scientists refract issues through their disciplinary and professional perspectives. The public agenda, partly set by officials, partly reflecting the actions of scientists and other groups and always evolving in a dynamic tug of interests, does not completely mesh with the agenda of narrower concerns defined by scientists. How these fit together or fail to fit provides the focus of the debate on science policy.

The issues now of greatest concern to the nation relate to industrial competitiveness, trade, and economic performance. The nation needs to rebuild and to modernize its industrial plant, reclaiming (or maintaining) leadership in technological innovation and strengthening its competitive position in the new world economy. In the light of lessening East-West tensions, critical choices have also arisen over whether to pursue technological innovations in new weapons systems as the most secure path to stability, or to negotiate further arms control and troop reduction agreements with the Soviet Union. As defense budgets decrease, should the technology base underlying the nation's security be protected from cutbacks or should force readiness be stressed even at some cost in modernization? Competitiveness and national security overlap in complex ways. It is a matter of vigorous debate whether the defense R&D effort helps or hurts American economic competitiveness. Do advances in defense technologies transfer into the marketplace? How important to national defense are controls preventing the export of sensitive technologies, particularly when changes in the Soviet Union and Eastern Europe seem to have diminished dramatically the military threat?

Beyond these interconnected issues are lesser but still important concerns that include managing the telecommunications revolution, controlling costs in high-technology medicine, confronting the ethical and practical consequences of prolonging life and of longer life spans, regulating public health and safety, finding clean long-term sources of energy, and controlling the climate and dealing with global warming. The total makes up an important part of the public policy agenda as viewed by politicians, the news media, interest groups, and concerned citizens.

Scientists, engineers, laboratory directors, university presidents, and others from the scientific community share many of these concerns, and as citizens help to articulate and shape the agenda. But they also worry about the health of their own enterprise and look for ways to strengthen and safeguard its future. There is no unified outlook, but they are concerned about replacing deteriorating facilities and instrumentation in research universities, attracting promising American students into the graduate engineering programs and dealing with future shortages of scientists, strengthening the linkage between academic and industrial research, keeping university research efforts in touch with rapid advances in industrial research, reducing poor public understanding of science, and improving the quality of scientific training in elementary and secondary schools. The increasing capital costs of research have made it harder for universities to maintain vital scientific work. Inadequate funding, creeping bureaucracy in the system of research grants, and growing politicization of the government-science relationship also worry scientists. The pork barrel distribution of scientific facilities particularly troubles those who worry that objective standards for allocating research grants are becoming increasingly difficult to maintain.

Industrial scientists seek renewed emphasis on improving manufacturing technologies and the production process as a route to competitive advantage. The neglect of production improvements and the resulting decline or relative loss of product quality in comparison with the advances of Japan and other nations is now widely recognized by industrial managers. They see accounting and financial criteria that emphasize quick returns and discourage long-term investments in new technologies as part of the problem. Regulatory changes in the economy, such as AT&T's divestiture of the operating companies, new policies in hospital cost reimbursement, and oil and gas deregulation, may also have far-reaching effects on industrial research.

Industrialists worry that excessive regulation could choke off research in biotechnology. Changes in policies allowing industrial research and commercial development under defense contracts could hurt aerospace and electronics, and erode the defense technology base; wrong moves in environmental regulation could harm the chemical, pharmaceutical, and related industries. The effects of tax reform, export controls, technology licensure and copyright agreements, and mergers and leveraged buyouts on American industry and its underlying technical capability are difficult to assess but are matters of concern.

Nonprofit research institutions, national laboratories, research divisions within government agencies, and those who operate "big science" facilities in various fields also have concerns arising from the desire to serve national goals more effectively and from the instinct to improve and protect their own institutions. As with all institutions it is easy to confuse self-interest with national interest.

Clearly, many of the goals defined by public policy converge with those from the institutions engaged in research. The nation can find common cause between what is good for all citizens and what is good for science and technology. Rebuilding industry, strengthening the universities, creating a more highly trained citizenry, and maintaining a strong defense based on technological leadership are efforts most Americans can cheerfully embrace. But just as clearly, differing opinions on achieving these goals and reconciling conflicts among them preclude any easy consensus, and the links of institutional and bureaucratic self-interest with the public interest are always difficult to establish. Science and technology can certainly help solve national problems, but deciding which parts of which problems are reasonably amenable to scientific and technical solutions is complex.

Many of the causes of the most pressing national problems cannot easily be fixed by technology, but there are few to which science and technology are wholly irrelevant. Finding effective strategies for the nation means sorting out how technical measures interact with macroeconomic policies, political leadership, customs and habits, and other factors to influence behavior at many different levels. Some problems may be resolved by changes in the behavior of companies, some through individual incentives, and others through government action; and many involve a mixture of governmental, institutional, and industrial responses. Since current problems on the national agenda can be traced to actions taken or not taken over many years, effective strategies will likely involve the same complex mixture of actions by the public and private sectors. Science policy is properly a part of this public debate. The task of this book is to show how the nation has tried to employ its scientific and technical resources to achieve important goals and to cast some light on ways it can do so more effectively. In particular, I seek to assess whether the nation has benefited from the loose pluralism of the research system and whether current policymaking should be centralized to better serve future needs; whether the universities, government labs, and industrial labs need to be strengthened in their own roles and in their interaction with one another; whether there is an unhealthy imbalance between defense and nondefense objectives, especially commercial ones; and whether the use of science in regulatory decisionmaking can be improved.

2 || Science and Government before World War II

THE ATTITUDES of the American public toward early American efforts in science and technology have traditionally been as simple and stylized as if drawn from a McGuffey reader. The triumph of American science was, since the beginning of the Republic, supposedly the product of unique and interrelated American traits. Americans, it has been thought, were indifferent toward basic research; what they admired was useful knowledge, as evinced in their infatuation with gadgets and inventions of all kinds. Leave the basic ideas, the fine-spun academic theories, and even the first crude technical applications to European science; America's strength lay in progressive improvements. These innovations were driven by commercial motives and market forces: "If a man . . . make a better mousetrap," Ralph Waldo Emerson said, "though he build his house in the woods, the world will make a beaten path to his door." A corollary therefore was skepticism about government intervention in developing new technology. How could bureaucrats know better than inventors the needs of the American people? Government funding would have meant government control and the kind of centralization that had stifled the commercial life of France and created the endless theorizing of Germany.

However familiar these images are in outline, they are in fact caricature. The reality of American scientific research and innovation before the twentieth century was far more diverse and its relationship with government funding and direction far more complex. The public iconography of a Benjamin Franklin, a Robert Fulton, an Eli Whitney, or a Thomas Edison emphasized the lone inventor, outside any scientific community, disdainful of theory, working on a practical problem, away from even the passing interest of government. Thus Franklin was remembered as flying a kite to prove what others had theorized—that lightning and electricity are identical—and as inventing bifocal lenses and the stove named after him: a sort of genius in homespun. But Franklin the theoretician who hypothesized and named the positive and negative poles of an electrical current in several widely circulated scientific papers, Franklin, a founding member of the American Philosophical Society, and Franklin the advocate of wider support for scientific endeavor have been left out of the public image.

There is then a significant disparity between the myth of American science and scientists before World War II and the reality of an innovative and flourishing community in which members kept informed about discoveries in America and abroad, nurtured theory and made important contributions in some disciplines, and from the first depended on government for at least some significant portion of research effort and support.

The Early Growth of Scientific Endeavor

American participation in the flowering of Western science in the late eighteenth and early nineteenth centuries created, in some fields, unexpectedly important contributions. Especially in zoology, botany, and paleontology, in which European studies lagged, America's naturalists provided experimental observations and specimens that revolutionized classification systems as well as enriching collections such as those in the British Museum. American scientists also made theoretical advances of a modest sort in astronomy, chemistry, geography and earth sciences, and comparative linguistics. Yet from the start American science showed distinctive practical traits, as the motto of the American Philosophical Society of Philadelphia, "the promotion of useful knowledge," makes clear. The preface to the first volume of the society's *Transactions* stated, "Knowledge is of little use, when confined to mere speculation. But when speculative truths are reduced to practice; when theories, grounded upon experiments, are applied to common purposes of life; and when, by these, agriculture is improved, trade enlarged, the arts of living made more easy and comfortable, and, of course, the increase and happiness of mankind promoted; knowledge then becomes really useful." [1]

The emphasis on useful knowledge was not confined to Benjamin Franklin, John Bartram, and their colleagues. The other large scientific association in the new country, the American Academy of Arts and Sciences founded in 1780 in Boston, also sought to encourage "useful experiments and improvements, whereby the interest and happiness of the rising empire may be essentially advanced." [2] Lacking the well-equipped universities that would promote a high level of scientific activity in the European pattern, Americans, perhaps only naturally, emphasized the experimental and the utilitarian and identified science closely with technology.

The Founders included many men well versed in science. Franklin was a scientist of world renown. Thomas Jefferson was an extraordinary enthusiast, patron, and promoter of scientific efforts in a wide variety of disciplines. As men of the Enlightenment, these leaders strongly believed that all knowledge would ultimately be practical and useful. Their conception linked learning to practical betterment and moral improvement, and the deism of many of them allowed a comfortable fit between science and religion. The advance of knowledge glorified God as it uplifted mankind and improved living conditions in the new republic.

The science encouraged by the American Philosophical Society was not, however, as clearly utilitarian as the preface to the first volume of the *Transactions* would demand. The rest of that volume was devoted to a discussion of the transit of Venus, which earned the respect of Europeans but did little to promote the useful arts in America. In fact, dissatisfaction grew over the Philosophical Society's insufficient practicality; some critics concluded that its focus was too broad to advance manufactures, agricultural improvements, and other special concerns. Thus a number of societies devoted to more specialized scientific inquiry arose, including, in Philadelphia, the Chemical Society, founded in 1792, and the Academy of Natural Sciences (1812); in Boston, the Massachusetts Society for Promoting Agriculture, Massachusetts General Hospital, and the Boston Athenaeum (all by 1807) and the Boston Academy of Natural History (1830); the South Carolina Society for Improving Agriculture (1785) and the Medical Society of South Carolina (1789); and in New York the Society for the Promotion of Agriculture, Arts, and Manufacture (1791).[3]

The Revolutionary War cut America off from close ties and cooperation with European science and gave a nationalistic impetus to the budding scientific effort. Moving the capital to Washington also helped ensure that no American city would dominate American science in the way London or Paris set the tone for England or France. Science here would be decentralized. Critics have argued that the Revolutionary War period, with its democratizing influence, led to American indifference to basic science during the nineteenth century.[4] High science in the European pattern, they have stated, was more congenial to the aristocratic and Enlightenment cast of mind displayed by American gentlemen amateurs before the Revolution. The more applied orientation became pronounced afterward, narrowing the focus of inquiry and pushing scientific enterprise toward a radical Baconianism mired in facts and hostile to theory. This bias supposedly dominated American science until the end of the nineteenth century, when professional scientists began to staff the universities, and private research institutions became more favorably disposed to research in pure science.

The view is misleading. In an authoritative study, Robert V. Bruce has shown that American science was a professional pursuit earlier than formerly imagined, was much more than a merely derivative branch of European science, and was related in a productive way to an emerging engineering profession.[5] A full review of these developments is beyond the scope of this chapter, but a few important points about nineteenth century American science deserve special attention.

Applied and Basic Science in the Nineteenth Century

Despite the doctrine of useful knowledge, there was little applied research of significance in the young republic: as John C. Greene states, early experiments soon showed that "the application of science to practical affairs was a long, slow, expensive business."[6] The nation's research "system" was neither sufficiently developed nor well enough integrated to allow scientific energies to be organized to promote concrete purposes. But about 1815 scientific activity of all sorts, including applied research within the federal and state governments, quickened. The need for mapping, surveying, defining standards, and designing ordnance gave rise to government technical bureaus that were increasingly useful. At the federal level most of these activities supported internal improvements or national defense: the nation's first engineering school was the Army Academy at West Point, New York, formally designated a center of instruction in 1818. The efforts could usually be classed as applied research, although directors of government technical bureaus in the nineteenth century were also part of the nation's scientific leadership. Alexander Dallas Bache of the U.S. Coastal Survey, Matthew Fontaine Maury of the Navy Department, and Joseph Henry of the semipublic Smithsonian Institution were particularly influential figures.[7] In collaboration with colleagues from the universities, Bache was instrumental in creating the National Academy of Sciences in 1863, a society for honoring the nation's outstanding scientists and a source of scientific advice for the federal government.[8]

Americans did not confine their scientific interests to any single discipline; they were concerned with chemistry and astronomy as well as with natural history and anthropology.[9] American science was, however, more empirical than French or German science and more preoccupied with observations. Scientific inquiry in America also conformed more to the style of the British than to the styles of the Continent. But any bias in favor of raw empiricism seems to have been discarded by the middle of the century.[10]

Despite their interest in research per se, however, scientists couched appeals for support in utilitarian terms and generally justified themselves to the public on the basis of practical contributions to society. As science became a more professional pursuit, they increasingly saw themselves as a distinct class with distinct interests, styles of communication, and standards of judgment. Decisions on scientific matters were made at successively lower levels in government agencies, and the guild consciousness of the scientists was correspondingly strengthened. What emerged was a pattern of simplified appeals for support made by scientists to politicians largely ignorant of science. This pattern differed sharply from that in the eighteenth century, when statesmen who were science enthusiasts engaged in spirited and informed debate on the idea of a national university, the patent clause, explorations, and other scientific concerns of the day.

Examining the divergence between the views of the new scientists and those of the politicians and public may clarify much of the scholarly controversy about whether America was indifferent to basic research in the nineteenth century. Alexis de Tocqueville's observations in the 1830s and those of later observers on the practical and antitheoretical bent of Americans seem accurate; but Tocqueville was not studying people engaged full time in the pursuit of science.[11] Thus historians who trace the emergence of the scientific profession in the nineteenth century and who find a strong commitment to basic science are also correct. The difference arises because these historians focus on the statesmen and practitioners of science. The attitudes of scientists simply differed from those of the rest of the public.

Appeals for public support, however, were still framed in terms of practical benefits, which promoted the misconception of the utilitarian character of scientific inquiry in America. Scientists exaggerated what they could deliver or misjudged the time it would take to produce results. Politicians knew this and made allowances for it. The impact of research efforts was far more profound than anyone foresaw, though usually the results came more slowly and less predictably than anticipated.

By the 1870s a large and important scientific community had emerged. Most of the major types of institutions—universities, government research laboratories, independent research centers—existed.[12] Edison's Menlo Park "invention factory" opened in 1876 (it was, however, not yet a modern industrial laboratory). Earlier, Yale had become the first American university to adopt the German practice of awarding the research doctorate. In the late 1860s Josiah Willard Gibbs, the first internationally important American theoretical scientist since Franklin, was awarded the third Ph.D. granted by Yale, which led in establishing pursuit of the doctorate and an emphasis on research as the model for graduate study. Research universities became the home of science in the United States.[13] American scientists engaged in fundamental research in all scientific disciplines. The work was supported (and performed in some instances) by the organized philanthropies created during the economic expansion that followed the Civil War.

The value of a domestic research effort in the basic sciences came to be acknowledged by scientists and the country's leaders, but few endorsed views that the government should support it. Most believed government support should be directed toward applied research to advance the government's missions. The more basic research of the universities and independent institutes should be generally funded by the institutions themselves and by organized philanthropy. The growth of the elite Eastern private research universities in the late nineteenth and the early twentieth centuries was, in fact, accomplished mostly through philanthropic support, which also helped develop such public institutions as the University of Chicago in those years.[14]

Many people also believed that the nation could import scientific ideas from Europe, just as it imported capital. While the United States was not dependent on Europe, a lively flow of ideas and frequent exchanges of people contributed significantly to the development of American science and technology. But successfully importing scientific ideas or technical processes required considerable technological sophistication, a skilled work force, and appropriate organizational and managerial capacities. To absorb foreign technology, America required commercial research activities of its own, and these were carried out in the design shops, the engineering departments, or increasingly in the industrial laboratories that evolved in big companies at the end of the century. This effort was, of course, financed by the companies themselves; it was not directed by any government master plan.

The Contrasting Pattern of Agricultural Research

A significant exception to the pattern of private funding was research in agriculture and its supporting disciplines. Here the nation attempted to pursue both scientific and commercial leadership and to improve the quality of life in rural America. Beginning with the Morrill Act in 1862 and continuing in other legislative and administrative actions, the federal government built an integrated system that included basic and applied research, agricultural experiment stations, county agents, and extension services to support the farmer. The Morrill Act had been preceded by abortive efforts from associations of prosperous farmers to support research privately, but because of the problem of free-riders who could benefit from the research without having to pay for it, these groups decided that their interests would be better met by federal programs that served all farmers.

How this all came about and how the system persisted until it had to face the twin challenges of economic deregulation and the revolution in the biological sciences in the late twentieth century is a story that needs to be told.[15] But generally because of a historic commitment to agricultural self-sufficiency, because all knowledge relating to soils, plants, pest control, and crop yields seemed directly useful, and because, once established, political interests perpetuated and expanded the system, the federal government consistently supported agricultural research for generations. It was as though the nation had chosen to hedge its faith in pluralism with a major experiment that ran counter to its belief otherwise in private support for research. The aim in this case was to link government, university, and producer interests closely to improve agricultural productivity at virtually any cost. Critics have pointed to the relative isolation of the agricultural research system from other scientific research as a cause of its failure to respond to the dynamic developments in the biological sciences.[16] But defenders have noted that other parts of the system, such as the State Technical Services Act of 1965, were from time to time modeled on the agricultural extension services. Geographical equity and the importance of a stable political constituency for research have been generally admired features of the agricultural system.

Nineteenth Century Development of Technology

America's supposed gift for innovations in technology is the obverse of its supposed indifference toward basic scientific research. From an early date Americans were thought to exhibit great practical skills, enabling them to invent new products such as the cotton gin or improve upon European technologies. With these special gifts, self-taught inventors and engineers, with no aid from formal science or the government, conquered a continent and launched the world into a new phase of the industrial revolution. Incremental advances in technology, fueled by the profit motive, led to rapid if uneven economic growth through most of the century. The situation supposedly lasted until companies began, toward the end of the nineteenth century, to institutionalize the process of innovation through research. This story needs revision.

Originally there was no indigenous American technology, if one means machines and other devices supporting European-style economic production, transportation, medicine, and the material conditions of life. (The word *technology* was coined by Jacob Bigelow in 1831.)[17] European settlers brought the physical contrivances of European civilization. The industrial revolution, starting with machines for weaving cotton textiles in Manchester and spreading to the rest of England from 1750 to 1800, transformed the production of goods by using a disciplined labor force, more labor-saving machinery, and higher capital outlays.[18] The industrial revolution may be said to have migrated to America in 1790 in the person of Samuel Slater. A skilled mechanic, Slater had memorized in detail the production process used in Manchester textile plants, then smuggled himself out of England in defiance of export control laws. After making his way to the wealthy Browne family of Providence, Rhode Island, he set out with them to launch the textile industry of North America.

In 1813 Francis Cabot Lowell, a self-taught engineer who also stole designs, plans, and factory layouts from England, began the first textile mill that combined both spinning and weaving—in effect, the first factory in North America. But then as now, technology could not simply be borrowed and immediately installed. Setting up production machinery and making it work takes time because ideas must be mastered and skills acquired. This acquired learning then ripples through the economy. In this fashion—through the adaptation and incorporation of borrowed technology—the industrial revolution began in America.

Meanwhile, Americans had invented devices ranging from dropleaf tables to apple corers to improve their daily lives. But these were not yet products in the formal sense because there were no markets in which they could be sold. Usage tended to be confined to the individual household or to a small locality. Improvements made early in the nineteenth century to plows and other agricultural implements were the first to find wide application. Regional markets thus began to develop, adumbrating the growth that would occur in the wake of the communications and transportation revolutions that occurred in the middle years of the century.

So far the familiar image approximates the reality: many small technical advances fed economic activity, technology was only marginally connected to basic research, and government support for innovation was very modest. But while Americans displayed technical ingenuity, their skills were no greater than those of other peoples. Technologies rapidly circulated around the world, albeit not with the velocity by which newly industrializing countries absorb them today. Nonetheless, international competition existed, and technologies were transferred from the centers of economic activity to peripheral areas. Again, the federal government's direct influence on industrial development may have been modest, but government did provide the laws and policies that enabled development to proceed. The Constitution's protection of the rights of inventors and early policies on standardizing weights and measures helped shape the evolution of nineteenth century industry. The sale of patent rights and the resolution of patent infringement suits in the courts provided a measure of orderliness in the disorderly process of business expansion, consolidation, and merger. State governments also assisted development by aiding the building of roads and turnpikes.

What further role did government play in promoting economic growth? One could pause to debate the efficacy of the Revolutionary era's import-limitation acts in promoting domestic manufacture, or the effectiveness of the early tariffs in nurturing infant industry, or the impact of Hamilton's *Report on Manufactures,* or the effects of internal improvements. But even if one skips over this early period and assumes that entrepreneurial energies were more important than public policies in the initial growth of the nation, disparities between history and myth multiply as one moves into the age of the telegraph, the railroad, the beginnings of standardized manufacture, and the growth of the iron industry. A clear overview becomes harder to provide. Samuel F. B. Morse, whose simple telegraph design came to dominate the industry in the United States, successfully appealed to Congress for a $30,000 grant; he wanted to build a line from Baltimore to Washington, a demonstration project he considered critical in launching the new technology.[19] Morse, in fact, saw government as the principal customer for his invention, just as scientists in 1940 envisaged the computer as a product best suited for government use. Government use of the telegraph was important to early growth of the industry, but that use then declined, which helps explain why the telegraph did not become a public monopoly as it did in Europe.

A federal grant to the Franklin Institute of Philadelphia in the 1830s to investigate the causes of steamboat boiler explosions contributed to the development of new materials for industry.[20] Procurement specifications by the army arsenals at Springfield, Massachusetts, and Harper's Ferry, Virginia, encouraged standardized manufacture in the woodworking and small arms industries, and these techniques slowly diffused into other sectors of the economy.[21] The army also loaned officers to conduct surveys for the western railroads.

At the same time, the federal government removed tariff protection for the American iron industry. This deregulation, to use a modern term, stimulated the industry to quicken its pace of technological innovation and catch up with the production techniques employed overseas.[22]

The government's role in the evolving industrial economy thus cannot be characterized by a few simple formulas. Other popular beliefs about the nature of American technological prowess also need revision. Morse, for example, was hardly a simple yeoman tinkerer. He had studied science, especially the principles of electromagnetism, at Yale and New York University as well as in Paris. (Erudite and cosmopolitan, he was also one of the leading portrait artists of America and had studied in Paris with European masters.) His critical idea for the early telegraph came while he was returning from Europe. And he was assisted in improving his early design and his experimental apparatus by Professor Joseph Henry of Princeton and Professor Leonard Dunnell Gale of New York University. Morse was himself a faculty member at the university (albeit a professor of design) at the time of his invention.

Notwithstanding his image in the popular imagination, Thomas Alva Edison does not fit the stereotype either. Although he had little formal education, his knowledge of the theory of electricity was exceptional, and his lengthy apprenticeship in the telegraph business enabled him to acquire a thorough knowledge of the technology, the materials, and the operations of the high-technology industry of the day. His early inventions were improvements on various instruments of telegraphy and were based on what would now be called a meticulous systems engineering approach.

Not only knowledgeable in matters of science, Edison was also a savvy entrepreneur. He chose to leapfrog the arc carbon lighting technology in favor of working with incandescent light, largely because the limitations of carbon arc lighting—it could only be used outside or in large, open interior spaces—meant that, even if improved significantly, it could never light homes or offices. Here, he reasoned, lay a potentially vast market. The invention of the incandescent electric lamp was not then the result of tinkering by an amateur in an informal, small-scale, low-cost fashion. As it was for Edison, much of the technological genius of America was a genius for organization, for management, for either linking an invention to a rapidly developing market or creating a new market for an invention. For example, the invention of a machine to produce over 120,000 cigarettes a day led, in one of the initial triumphs of marketing, to a massive effort to persuade nonsmoking Americans to smoke and to change people's dependence on pipes or homemade cigars or the ubiquitous chaw to a dependence on cigarettes.[23]

America's greatest invention of the nineteenth century was perhaps no single technical innovation but the creation of the large business enterprise itself and the capacity to manage it.[24] Technology, of course, was important in creating large, vertically integrated companies. But it was not mastery of technology as such that distinguished the American company nor technology that gave it the special place it occupies in creating a modern mass consumption society. The industrial revolution in Britain, and its spread across Europe with different organizational patterns during the nineteenth century, also featured companies built around the mastery of a new invention or a superior product or process, along with marketing skills. But the American company was marked by the successful linking of research or engineering with production, which was in turn supported by marketing, finance, and management control.

As rapidly as the modern business enterprise emerged in America, however, changes began occurring that implied trouble for the future. Manufacturing gradually became more routinized and less important in the organizational hierarchy. In place of the inventor who started the company, professional managers with specializations in finance or marketing often became chief executive officers of the maturing corporation.[25] Invention was taken for granted. Within the technical operations of the company a further stratification relegated production to the position of lowest prestige. The design engineer retained a status that the engineer engaged in manufacturing—a blue-collar function—lost. The industrial research laboratory became the home of the researchers and designers, not an aid to manufacturing.[26]

Although America seemed to have found the formula for creating a stream of new products and maintaining continuous growth, there were problems in making this more complex form of business enterprise function smoothly. The research laboratory tended to become absorbed with research and technology as ends in themselves. Management found it increasingly difficult to identify at what point technological innovation could be translated into a product with commercial potential. A design that worked in the laboratory did not always work when the company attempted to manufacture the product in volume. Meanwhile, production, as specified in theories of scientific management, became a mind-numbing, repetitive task, often defined in excruciating detail and broken down into suboperations performed by workers isolated from each other and divorced from any creative role in carrying out assignments.

Moreover, the technological sophistication exhibited by the most advanced American firms was only thinly spread in the economy. Many companies had little capacity for original research or for designing their own process technologies. Their product designs were derivative and their technologies borrowed or purchased off the shelf. They produced goods that sold only because demand was high and buyers unsophisticated. These shortcomings were bound to become evident as industry developed.

To note these aspects of industrial development enables one to understand more precisely America's peculiar skills and weaknesses. The nation possessed a capacity for developing technology, but hardly one of any unique character. It was the effective management of innovation and its integration with the company's business operations that was critical. Achieving a slight advantage in production or marketing enabled many businesses to create and exploit dominant positions in rapidly expanding markets and to grow large enough to incorporate supply, production, and distribution within a single organization. This, and not some unique gift for technology, was the American genius. The system was particularly well suited to a period of relatively stable process technologies and long product cycles. These conditions existed even after World War II: large expenditures on fixed capital assets and tooling were justified for high-volume production of standardized goods. But this system was less suited to a period of rapid change in technologies, short product cycles, and one in which a better-educated consumer expected high quality and reliability.

The Research System at the Turn of the Century

By the first quarter of the twentieth century, the nation's research system had reached approximately its modern form. Major institutions had become fully articulated performers of research and technical services and had a stable pattern of interaction. A rough division of labor existed for basic and applied research, development, and production. Industrialists, government scientists, and university leaders generally found the arrangements satisfactory. It is too much to speak of the emerging relationships as based on settled policy or explicit understanding, which would imply a greater mutual accommodation and a crisper definition of roles than any of the participants would have accepted. Instead, the system was loosely integrated, with some overlapping of roles and no central direction.[27] Above all, the government did not guide the enterprise. The leaders of basic science generally resisted government support because they feared the corrupting effects that populist politics might have on scientific inquiry, a position they maintained until the Depression created rifts within their ranks.

Within government the agencies and bureaus took over direction of scientific and technical efforts. Except for an abortive attempt in the early 1880s by a small group of scientists and their congressional allies to create a Department of Science, there was no serious challenge to the idea that each agency would conduct such research as it deemed useful to accomplish its missions.[28]

Government technical work was rarely contracted out (the Civil War years had been an exception). These activities proceeded separately but were never rigidly divorced from industry. Improved standards, carefully defined weights and measures, and precise technical specifications accompanying procurement were modestly helpful to industry. The emerging elements of the regulatory state in its technical aspects—food and drug regulations, safety standards for construction, and so forth—were a mild annoyance to industry.[29]

Industrial research was decentralized and competitive, following the structure of the economy. The merger of finance and industrial capitalism sought, for example, by J. Pierpont Morgan (which became the pattern in Germany and Japan) had been more or less blocked by the antitrust movement. Large, vertically integrated firms dominated many parts of the economy, but competition among regions, sectors, and product categories continued, and attempts to exploit new technologies guaranteed rivalry and dynamic change. Still, only large companies could afford well-staffed laboratories and extensive product development operations. Technology was thus widely but thinly spread. Industry-university technical contacts took various forms, reflecting no standard model and certainly no unified direction or overall goal. On the eve of World War I, most of the pieces that make up today's research system were in place, but interactions were relatively weak.

Beginnings of Modern Science Policy

On the eve of World War I, the applied research activities of the military departments were in a state of neglect. While government research efforts linked to mining, agriculture, geologic surveying, weather prediction, and public health flourished, the military had allowed the efforts to use scientific advice begun during the Civil War to lapse (the efforts to transform the army from a state-based militia to a national institution had consumed much of the army's attention).[30] Only the Signal Corps had shown imagination and energy in using the results of early research on radio communications. The navy supported the Naval Observatory, had made the transition to modern ship design, changed from coal to oil as a fuel for ships, and was operating with improved and modernized gunnery practices. But these changes seldom proceeded from the navy's research. They were imitations of the latest European practices.

Such new weapons as planes, tanks, and poison gas, and industrial strength in such war-related technologies as chemicals, optics, and metals, seemed bound to play a critical role even as Europe became mired in trench warfare. The navy was the first to seek organized scientific advice and attempt to create a central research organization. On July 7, 1915, Secretary Josephus Daniels appointed the Navy Consulting Board under the chairmanship of Thomas Edison; eleven other members were designated by the nation's leading engineering societies. The board recommended the creation of a naval research laboratory (finally established in 1923), but mostly devoted itself to reviewing proposals sent in by civilian inventors to aid the war effort. Some 100,000 inventions were submitted and reviewed, but only 38 were found to have significant technical merit and only 1 went into production.

The National Academy of Sciences, which had no part in the creation of the Naval Consulting Board, was important in the broader mobilization of scientists and engineers for the war effort. Working through the Academy's National Research Council, George Ellery Hale and his colleagues assisted the War Production Board to recruit scientists into key posts as commissioned officers. The NRC also assisted in developing manufacturing processes for nitrates and optics, activating the universities as training centers, and coordinating technical efforts among the various wartime agencies.

A clear by-product of the wartime effort was the impetus it gave to industrial research activities generally. The ability of scientific research to produce new products and processes for use in the war meant that henceforth it would be fully exploited in peacetime. Industrial research thus became institutionalized in many more companies; laboratories mushroomed from some 300 in 1920 to 1,000 in 1927 and more than 1,400 in 1930.[31] The universities' contribution to the war effort was the Students Army Training Corps. Campuses across the country altered curricula, acted as induction centers for the armed services, and held drills and marching sessions. The SATC effort was, however, mercifully short-lived; it was demobilized two weeks after the signing of the armistice.[32]

Most of the wartime machinery for coordinating the efforts of government, industry, research institutes, and universities was dismantled soon after the end of hostilities. Although the war symbolized the marriage of pure and applied science in the nation's service, few permanent institutional traces of the collaborative effort remained afterward. The experience of the National Research Council is illustrative. The NRC failed to remain the central coordinating agency for science because the leaders of the scientific establishment were unwilling to concede the need for or the desirability of such a role in peacetime. They did not want science to become dependent on federal funds (even during the war the council had operated largely with private funds). The NRC did, however, achieve a permanent status within the National Academy of Sciences through an executive order of May 11, 1918, and played a modest role in the 1920s in promoting international scientific contacts, providing fellowships, and collecting scientific and technical information. Support for the activities came largely from the Rockefeller Foundation.

The major force in matters of science policy in the 1920s was Secretary of Commerce Herbert Hoover. A trained engineer, Hoover enlarged the scope of the Commerce Department's research activities, acquiring the Patent Office and the Bureau of Mines, and led it to provide technical support for regulating the aviation and radio industries. His most ambitious effort, however, was the attempt to create a national fund to support basic research in the universities. The attempt ultimately failed, but it reflected a notable advance in the level of thought about the functions of science and of the research system in American society.

Hoover decided that the research system was in trouble because universities lacked an adequate base of support. Although research efforts, largely applied, of both government and industry were well funded, universities had to depend on philanthropy, which was no longer enough to sustain their growing needs. If the basic research the universities conducted did not keep pace, applied research in industry and government would also eventually suffer. Three solutions presented themselves: government support, support from business, or general public philanthropy. Government, Hoover believed, should not confine itself to supporting applied research and should "accept the fact that the enlargement of our stock is no less an obligation of the state than its transmission." [33] While philanthropy should strive to provide more liberal support, the best answer lay in greater support from industry. Hoover accordingly organized industrial leaders and scientists prominent in World War I planning, along with other notables, to form the National Research Fund. They were to raise $20 million from industry to support basic research in the universities for a ten-year period. But though it was launched with great fanfare, the fund fizzled even before the onset of the Depression destroyed further hope of bringing in significant money. It spent no money on university research, and the meager sums raised were returned to the contributors.

Appeals for support from the fund's backers emphasized the inferiority of America's efforts in basic research compared with Europe's, but at least one member of the organizing group, J. McKean Carrell, was skeptical. He urged that a minute portion of the money the board proposed to collect be spent in "determining whether the first statements that it makes are correct." [34] This is one of the earliest calls for the study of science policy, and Carrell should be remembered for his pioneering effort to separate hyperbole from reality, a continuing challenge for those who toil in this intellectual vineyard. In fact, there were native-born and trained Americans who were as accomplished as any scientists in the world in such fields as theoretical physics and chemistry, including Edwin Kemble, John Van Vleck, John Slater, Robert Millikan, and Linus Pauling.

The nation's anticipatory effort to develop a policy for using and supporting science and technology was completed during the Depression, but the path toward a realized policy during the New Deal was neither smooth nor straight. Initially, science affairs did not fit comfortably with the social orientation of the New Deal. In his reelection bid, President Hoover had boldly asserted that science was a way out of the Depression.

His defeat seemed to discredit the idea. And in fact the mood under the new Democratic administration was unfavorable for science for this and other reasons.[35] Excess production and expansion that was too rapid were commonly considered the causes of unemployment. The codes of the National Recovery Administration, for example, stressed production limits to be negotiated industry by industry. The idea of expansion through innovation was distinctly unpopular. Technological unemployment was a very common explanation for the Depression, and there were calls for a moratorium on scientific research and technological development.[36]

Nonetheless the hard times facing the universities finally caused the ranks of the scientific leadership to split over the issue of government support for science. A new group of academic scientists (or in some cases a switch in party affiliation from Republican to Democratic by existing leaders) meant a more sympathetic attitude toward the New Deal and more backing for the idea of government support for university research, which they considered essential for the survival of their institutions. Scientists accordingly looked for ways to become engaged in helping the economy recover. An opportunity came when Isaiah Bowman, a geographer from Johns Hopkins University and chairman of the National Research Council, received an invitation from Secretary of Agriculture Henry Wallace to study the Weather Bureau. Bowman made a counterproposal for a broad Science Advisory Board with a wider mandate, a proposal that President Roosevelt adopted by executive order in July 1933.

With Karl T. Compton, president of the Massachusetts Institute of Technology, as chairman, the board studied a number of government science agencies during its two-and-one-half-year life. More ambitiously, Compton twice mounted a campaign to create the machinery and the funding to support university research on a significant scale. The first plan was a "recovery program for science progress" submitted to Secretary of the Interior Harold L. Ickes in September 1933. The plan called for using funds from the Public Works Administration to pay for university research on conservation and on the creation of new industries. It also called for building university facilities as public works, and for providing jobs for out-of-work scientists and engineers. Ickes was said to be "99 percent convinced that something of the sort should be done," but he concluded that "there was unfortunately no provision under the law whereby public works funds could be expended for research but only for construction."[37] Later Compton launched an even more ambitious plan for "Putting Science to Work!" This time he managed to get a watered-down version approved by President Roosevelt who, however, doomed the whole effort by adding the crippling stipulation that "90 percent of the funds expended must go to direct labor paid to persons taken from the relief roles."[38]

The Science Advisory Board was absorbed into the National Research Council in December 1935. In deciding not to extend the board's independent existence, Roosevelt transferred its planning functions to the National Resources Board. Whether this board was a failure or a success, it was certainly plagued by divisions within its ranks. Influential social scientists in the Roosevelt administration considered Compton a representative of the physical sciences and hence unable to speak for the social sciences or for the institutional interests of higher education generally. Medical scientists were initially left off the board but muscled their way on to protect the interests of the nascent research operation of the National Institutes of Health.[39] The cause of the board's failure, however, was not so much internal friction as doubts among New Dealers as to whether technology was the source of the nation's problems or their solution.

Another effort in the 1930s that failed was the National Bureau of Standard's attempt to win Congressional approval for a program of basic research to be performed on behalf of American industry. The proponents evidently had in mind something akin to what would later be called mission-oriented basic research, but "the failure to define 'basic' adequately left the measure open, on one side, to charges of ivory-tower dreaming and, on the other, to fears that manufacturing processes would be patented and licensed by the Department of Commerce."[40]

Amid doubts and policy zigs and zags, the peacetime New Deal continued to nurture scientific activities and to move cautiously toward adopting the features that would characterize the research system after World War II. The National Cancer Institute was created in 1937 as the pieces of the modern biomedical research enterprise began to be assembled. By the late 1930s the Works Progress Administration routinely awarded research contracts to universities, and WPA-sponsored projects produced tables of mathematical functions that proved highly useful to scientists during the war. In 1941 the National Resources Planning Board, successor agency to the National Resources Board, put out a remarkable report, *Research—A National Resource*, that was a precursor to the Bush report at the end of the war.

So, on the brink of war the elements that were to be reconfigured to form the postwar research system were at hand if not yet firmly in place. A burgeoning academic research effort survived the Depression. Industrial research was restored to a strong position after severe budget cuts in the early 1930s. Active interest groups emerged to represent different disciplines as scientists abandoned their traditional political apathy. Strong research activities continued to serve various government functions (including regulation), and numerous planning and advisory units were in place. Except for a slight dip between 1932 and 1934, at the depth of the Depression following the banking crisis, the total national research effort increased each year in real terms between 1930 and 1940.[41] The orchestra had been assembled, but there was as yet no common score for the musicians and no conductor.

The war provided the transforming events from which the modern research system emerged. The most dramatic events were those leading to the creation of the Mahattan project, in which scientists played a leading role from the extraordinary letter signed by Albert Einstein and others, alerting President Roosevelt to the dangers of atomic energy, to the final development of the first fission bombs.[42] In addition, scientific activities in general were raised to a more prominent position in the defense agencies. The military services expanded their laboratories and extended their contractual relationships with industry and the universities. The practice of relying on industry for development and production and on universities for research was clearly established. The universities were fully mobilized for war through the creation of large campus laboratories such as the Radiation and Instrumentation Laboratories at MIT or through university scientists joining in direct government service for the duration of the emergency. Normal scientific progress was interrupted, but new fields of applied science and new developments in technology were stimulated.

At the White House, the Office of Scientific Research and Development under the leadership of Vannevar Bush served as the president's agent in setting policy and coordinating the mobilization of the research effort.[43] Many practices—the research contracts with universities, the growing support of mission-related research by government agencies, the enhanced status for scientific advisers and technical bureaus—took shape and carried over into the postwar world. But wartime practice by no means dictated a clear position to be followed later.

The emergency, indeed, barely suppressed sharp quarrels about the function of science in American government and how society should support science. In the ideological crosscurrents of the New Deal, many populist elements continued to be influential in the early postwar period. Typified during the war years by Senator Harley Kilgore of West Virginia, the populists tended to combine strong views on patent law reform with a preference for formula-type research funding. This would, in their view, help prevent domination of the research system by the big private universities, presumed to be in unholy alliance with big business.

The wartime scientific leadership, while not wholly united, strongly disputed the populist tendencies. Bush, depressed by bureaucratic battles with the military and concerned that creative research would be stifled if traditional procurement practices persisted after the war, wanted the wartime machinery to be dismantled quickly and new practices adopted for peacetime. Formula funding and domination of the research system by the big government departments were to be avoided. Instead he thought an independent agency should be established to operate more flexibly. Science had, however, clearly shown its usefulness to the nation, and a close partnership of some kind between government and science was bound to emerge after the war.

3 ‖ The Postwar Consensus, 1945–1965

THE EXPERIENCE of World War II was the decisive event transforming the nation's approach to science and technology. The war and the revolutionary developments growing out of it made policy for science seem necessary. The time had come to think about what large-scale scientific research meant for American society and democracy. The Priestleyan vision of a self-governing science driving the world toward benevolent ends could no longer guide affairs. What should be the structure of postwar support for science? What implications for the economy would grow out of wartime scientific experience? These were only a few of the urgent concerns that emerged from developments in atomic energy, space science, information technologies, and, a little later, biological sciences.

In response to deliberations over mobilization for war and then of demobilization followed by the semipeace of great-power rivalry came the postwar American consensus on science policy. There are no sacred scrolls revealing the articles of faith. The Bush report, *Science—The Endless Frontier*, perhaps comes closest to summarizing many elements of the consensus, but even it is incomplete. Rather, one must reconstruct the consensus from many sources, among them the legislative histories of the National Science Foundation and the Atomic Energy Commission, the conclusions of the Steelman and other postwar reports, the speeches of principals, and the deliberations of advisory bodies. But the consensus is no less real and critical in its impact for that.

Among the major tenets was that basic research drove the system, providing the advances that would sustain the pace of inventions and applications. And since the benefits would be too far in the future for industrial support and the costs too great for philanthropy alone, the government would have to assume responsibility for nurturing the effort, which would take place primarily but not exclusively in the universities.

Applied research was to be an important government responsibility as well, but would be more fully developed and better articulated with other research. Each agency was to strengthen its network of supporting technical bureaus and relate the missions of those bureaus to each other and to any contract research done by universities or development work by industry. Federal research efforts would be concentrated most heavily in the large departments or agencies in which the government itself was the purchaser or consumer of the end product. The agency mission might require interest in a wide range of technological systems and their supporting sciences, as with the Defense Department, or might be more closely related to the progress of a narrower range of technologies or scientific fields. Since science and technology were erasing many jurisdictional boundaries, every government agency would increasingly look for new missions or for new ways to perform old ones.

Commercialization was to occur almost automatically as a by-product from the government's support of basic research and more applied research and development operations. The incentives of the marketplace would continue to spur innovative activity, and the manufacturing strength of U.S. industry would ensure high-quality products and quick exploitation of market opportunities. Trained personnel produced by the universities and the start-up markets provided for some high-technology goods and services via government procurement were to give an additional boost to the workings of the traditional market system. So the best policy on commercialization was to have no policy at all. Industry would develop products and know best how to commercialize technology. The fact that U.S. manufacturing productivity (table 3-1) was more than twice that of its principal competitors certainly lent plausibility to the idea.

Regulation would be required for a few new services and products, as it had been before with railroads, drugs, and radio, but would be light and with only enough legislative framework to ensure public confidence. In general, the march of technology was considered benevolent and liberating. In those few instances in which undesirable side effects could be ascertained, the necessary minimun steps would be taken to render technology safe or to ban it. A related assumption was that there was ample time and opportunity to find a solution before much harm could be done. This inadequate recognition of adverse side effects from technological applications was understandable at the time because applications followed slowly from discoveries and the diffusion of an initial technological application usually took years.[1]

The foreign policy and national security dimensions of science policy followed straightforwardly from the above: America would rely on high technology to provide a strong national defense and to compensate for

Table 3-1. **Average Labor Productivity, Aggregate and Manufacturing, in Four Countries Relative to the United States, Selected Years, 1950–86**

Year	France	Germany	Japan	Britain	United States
AGGREGATE ECONOMY (GDP PER HOUR WORKED)					
1950	39.6	31.5	13.6	55.9	100
1960	47.5	47.9	18.4	54.8	100
1973	71.4	68.7	42.6	65.8	100
1980	85.8	83.7	51.5	71.7	100
1984	95.3	86.1	54.9	78.4	100
MANUFACTURING OUTPUT PER HOUR					
1950	32.9	28.6	11.4	36.3	100
1960	43.8	48.1	23.2	36.6	100
1973	65.8	66.8	55.3	41.9	100
1980	82.6	79.6	75.2	41.1	100
1984	85.8	78.7	81.3	45.5	100
1985	85.0	78.7	83.6	45.2	100
1986	83.6	77.5	83.0	44.9	100

Source: Martin Neil Baily and Alok K. Chakrabarti, *Innovation and the Productivity Crisis* (Brookings, 1988), p. 9.

any disadvantages it or its allies might have in numbers of troops or weapons. Technology would undergird the economy and national defense alike, though commercial and military activities would be more or less separate. Free trade would give ample scope for American companies to establish markets for their products, and they would stay comfortably ahead of the competition through mastery of technology. America would, however, sit tight on military technology and employ strict export controls to prevent transfer except to allies. If there were some leakage of military technology or if a foreign competitor acquired commercial secrets by unfair means, the annoyance would be unlikely to erode American security or technological leadership.

These ideas were loosely held together by faith in scientific progress. Joseph Priestley would have felt comfortable that the ideas he had exchanged with his friends Jefferson and Franklin were alive in spirit. And if the ideas were not wholly consistent and could not be said to form a tight philosophical system (what set of ideas in American public life ever met such a test?), their appeal was remarkably broad based. Leaders from industry, universities, executive-branch science agencies, Congress, political parties, news media, and the attentive public subscribed to them.

Growth provided the glue for the system. People supported the consensus in part because all shared the benefits of expansion. Federal sup-

Figure 3-1. **Federal, Private, and Total R&D Expenditures, 1960–84**

Billions of 1972 dollars

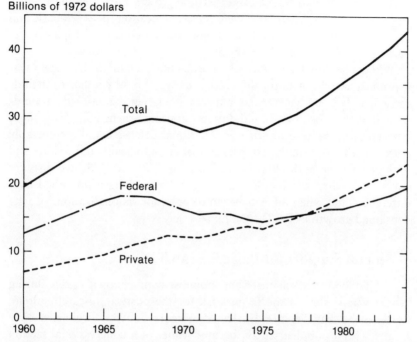

Source: National Science Foundation, *National Patterns of Science and Technology Resources* (1984), table 5.

port for R&D grew 14 percent annually in constant dollars between 1953 and 1961. The 1957 Sputnik launches ensured increased R&D spending for a decade afterward (figure 3-1).[2] The growth of federal research support served everybody's interests—universities, government agencies, industry, congressional committees—and helped blur inconsistencies and points of disagreement. Growth was both a condition and a part of the doctrine, and the two reinforced one another.

The research system was expected to grow rapidly and to continue to grow. The Bush report, *Science—The Endless Frontier,* had suggested funding guidelines for the proposed National Science Foundation, but these gave way to the more general recommendation of the 1947 Steelman report that R&D funding should amount to 1 percent of GNP.[3] This recommendation was later replaced with the guideline that a nation should strive for R&D spending from both public and private sources of about 3 percent of GNP.[4] Like a tithe, the figure was a goal to strive for, even if it were never quite attained. An alternative, reflecting the same spirit of expansion, was that 15 percent annual growth (a doubling time of five years) should be the target.[5] These growth rates were never matters of precise agreement, nor were precise allocations between basic and applied research; the matters were not considered fruitful topics of debate in the years of rapid expansion.

While a healthy basic research effort as the linchpin of the system was a primary article of faith, the consensus was broad enough to include those who felt that basic research deserved a larger share of federal funds as well as those who harbored doubts about its benefits. Similarly, those who stressed the role of science in national defense could comfortably coexist with those emphasizing its contribution to health and welfare and with believers in both "big science" and "little science." Heated disputes came later when the momentum of growth no longer smoothed over differences and when officials began to press for rationalizations of why continued growth in R&D funding was necessary.

Federal Support for Basic Research

A fight between scientists and populist congressional critics during World War II shaped the framework for the postwar research system. Early in the war, Senator Harley M. Kilgore of West Virginia, a New Dealer with a populist streak, became leader of a congressional faction dissatisfied with the performance of the war mobilization machinery and its alleged domination by big business.[6] Kilgore's views corresponded closely with those of Maury Maverick, an outspoken Texas liberal and member of the War Production Board, who wanted to break the grip of big business on the technical aspects of war production. In 1942, Maverick sought to persuade Donald Nelson, head of the War Production Board to create a new agency patterned along the lines of the Office of Scientific Research and Development to oversee production of strategic materials. Spurred by reports in mid-August 1942 that Nelson was about to move, Kilgore decided to preempt executive action by introducing the Technology Mobilization Act, a measure that would establish a superagency, the Office of Technology Mobilization, to direct all the government's technical bureaus, make patents and production processes available to small business, and remove bottlenecks to effective mobilization.

The bill encountered fierce opposition. It was attacked as a strange combination of regimentation from the top, giving government almost dictatorial powers to usurp patent rights and commandeer technical personnel, and a romantic belief in the powers of small enterprises and amateur inventors. Vannevar Bush, director of the Office of Scientific Research and Development, argued that such a radical step would undermine the mobilization effort and create rather than remove bottlenecks. The military services, the major companies engaged in defense contracting, and even former congressional allies assailed the measure. During extensive hearings in the fall of 1942, Kilgore modified his views and incorporated them in the Science Mobilization Act of 1943, a measure that sharply reduced the powers of the coordinating mechanism and removed many of the features considered most objectionable by critics of the first bill. In spirit and substance, however, the new bill retained its predecessor's orientation in favor of small business and strong public control of science.

Specifically, the proposed coordinating agency would be authorized to make grants to support scientific and technical education and to aid basic and applied research. Government support was important because Kilgore believed universities were unduly dependent on support from industry, which had reduced university research to "the status of handmaiden for corporate or industrial research, and has resulted in corporate control of many of our schools."[7] The agency would be administered with the assistance of scientists, who would serve on a board, and an advisory committee, but most board members and the staff would be nonscientists. Kilgore believed that control by nonspecialists would ensure effective representation of the public interest and keep the effort directed toward relevant ends. This provision began to define the major lines of the postwar battle over how to organize federally supported research.

As industry geared up for massive wartime production, provisions of the bill seemed to be overtaken by events. Rubber production, for example, had been one of Kilgores's early concerns, but by the end of 1943, tons of synthetic rubber were being produced. The threat of bottlenecks had largely passed. OSRD's Bush now urged the senator to forgo legislation on war mobilization and look ahead to the peacetime system for coordinating and supporting science. Thus in early 1944 Kilgore and his staff drafted a bill for the postwar promotion of pure and applied research and for technical education under a national science foundation. The bill again represented an attempt to deal with earlier criticisms, but on the fundamental issues the differences between the senator and most of the scientific leadership remained.

Kilgore envisioned a postwar structure under the control of nonscientists and directed toward practical ends. Most of the national effort would consist of strengthening government research organizations; they would have priority in receiving funds. Universities would also receive funds for research and technical education, but their role would be to support the larger national objectives advanced most directly by the government agencies. Most scientific leaders, however, believed the major postwar need would be to establish a clear policy for the support of basic research as a condition for a healthy national scientific enterprise and the basis for wider applications of research to national needs. This would mean special attention to universities and other nonprofit research institutions as the natural homes for basic research and producers of the nation's technical work force. (Government research laboratories could also be strengthened but should be limited to their traditional focus on narrower investigations of short-term operational matters.) Moreover, administration should be removed from politics and should not be subordinated to short-run pressures, which might defeat the larger ends sought. Insulated from such pressures, scientists would know best how to spend funds wisely.

Bush and other science leaders by no means disagreed wholly with Kilgore, and in the debate that had been under way for two years the two sides had moved closer. But no amount of politesse and surface accommodation could disguise their fundamental differences. Kilgore, in effect, sought to extend the agricultural research system to the nation's entire scientific effort (though he would have a central coordinating and funding agency). Bush and his colleagues would revive the spirit of Herbert Hoover's effort in the 1920s to promote basic research but would add public money—essentially without public control. The stage was set for a showdown, and Kilgore's early efforts had given him an edge in defining the terms of the debate. He had significant support from the New Deal's liberal wing, including Vice President Henry Wallace and trustbuster Thurman Arnold. It was in this context that the idea for a report to the president on postwar science affairs arose.

In October 1944 Oscar S. Cox, deputy administrator of the Foreign Economic Administration and an influential figure in the Roosevelt administration, came away from a conversation with presidential adviser Harry Hopkins with an idea to have Roosevelt appoint a committee to prepare a report on science. Though originally a campaign idea to show how the administration was seeking to use the results of wartime research to promote prosperity and full employment in peacetime, the concept was eagerly seized upon by Bush when Cox discussed it with him and OSRD General Counsel Oscar M. Ruebhausen as a way to regain the initiative from Congress in the debate on postwar science policy.

The result was *Science—The Endless Frontier*, presented to President Truman in July 1945.[8] The thirty-four-page main report, written by Bush, was accompanied by four technical appendixes corresponding to the four issues posed in Roosevelt's November 17, 1944, letter instructing Bush to undertake the review. It proved to be one of the most influential policy documents in the nation's history. Its ideas were not adopted in toto; indeed, the views expressed in the various appendixes were not wholly consistent. But the report succeeded in deflecting the ideas advocated in the Kilgore bill and in setting the tone and boundaries of the debate on science policy for nearly a generation. The research system that resulted incorporated elements drawn from the report and the Kilgore bill as well as from practical accommodations to postwar events.

The most important principle, announced in the report's opening paragraphs, was the fundamental importance of basic research to achieving vital national goals:

Progress in the war against disease depends upon a flow of new products, new scientific knowledge. New products, new industries, and more jobs require continuous additions to knowledge of the laws of nature, and the application of that knowledge to practical purposes. Similarly, our defense against aggression demands new knowledge so that we can develop new and improved weapons . . . [because] without scientific progress no amount of achievement in other directions can insure our health, prosperity, and security as a nation in the modern world.[9]

The report described the relationship between basic research and applications by using the metaphor of a bank or common fund from which deposits may be withdrawn: "Basic research leads to new knowledge. It provides scientific capital. It creates the fund from which the practical applications of knowledge must be drawn. New products and new processes do not appear full-blown. They are founded on new principles and new conceptions, which in turn are painstakingly developed by research in the purest realms of science." From this it followed that America could no longer rely on importing knowledge but had to make strong efforts in basic research a centerpiece of national policy.

Today, it is truer than ever that basic research is the pacemaker of technological progress. In the nineteenth century, Yankee mechanical ingenuity, building largely upon the basic discoveries of European scientists, could greatly advance the technical arts. Now the situation is different. *A nation which depends upon others for its new basic scientific knowledge will be slow in its industrial progress and weak in its competitive position in world trade, regardless of its mechanical skill.*[10]

This theme runs through major reports for the next two decades, including those of the President's Scientific Research Board (1947—the Steelman report), the President's Science Advisory Committee (1958, 1960), the Bell Committee on Government Contracting (1962), and the Committee on Science and Public Policy (COSPUP) of the National Academy of Sciences (1965).[11] As a corollary the need for the federal goverment to support both research and the development of scientists and technicians was nearly universally accepted. The technical committee assigned by Bush to look into postwar research funding and the public welfare did not at first clearly accept the need for federal support, but as the evidence accumulated that university research was decimated by the war effort, the committee strongly agreed that the need was there. It followed also that universities would be important participants in, if not the exclusive performers of, basic research. Their forte was free inquiry, and the need for scientific personnel (produced as a by-product of university research) could only be met through expanding academic research efforts.

How were funds to be channeled to the universities? The technical committee did not resolve the issue. Its model seemed to be the wartime administrative contracts between the Office of Scientific Research and Development and universities, but with full control in the hands of university officials. Funds, the committee suggested, should first be allocated to qualifying institutions automatically, matching the universities' own contributions to research. These universities would then make full use of the "accumulated wisdom of administrative officers and faculties" in internal distribution, thus saving the government from "the burden of investigating intensively the large number of potential recipients and arriving at a decision in regard to the merits and defects of each." At the same time, recognizing that such a procedure would mostly benefit the established private universities, which all had significant research programs, the committee urged that discretionary funds also be made available to research efforts in institutions with smaller programs that would not do well under the matching requirement. This would promote "a better balance of research activity throughout the country."[12]

Bush knew that such proposals would not be easily accepted by politicians. The idea of matching grants might even lead to more congressional support for the formula concept favored by Senator Kilgore whereby funds would be distributed to the states on the model of the agricultural research system. Without resolving the issue, the Bush report stressed the importance of discretionary funding by the proposed foundation, suggested reviewing the quality of research done in the universities, and made clear that contracts and grants could go to institutions with "talent" or those showing "promise" even if they lacked demonstrated research capacity. Assessing the quality of an institution so it could qualify for a matching grant might seem feasible if the number of research universities remained small, but what if, as seemed to follow from the objective that the nation's scientific resources should grow, the number of potential qualifiers were to increase sharply? Attention would thus appear to shift to the support of programs rather than institutions. Could this be done without control over university activities by their supporting agencies? The main Bush report and the technical appendix supplied by the committee were both at pains to emphasize that funding had to be stable, traditional procurement policies modified, and other steps taken that would protect university autonomy. Even so, it was apparent that "no matter on what conditions money is given to universities, the very existence of such support will, of course, modify university policy." [13]

It was finally only through the postwar legislative battle over the governance of the proposed National Science Foundation, and in the experience of the agencies that provided research funding to the universities while the battle raged, that a working answer to the problem was found. The successful formula amounted to assembling three elements into a coherent package: quality review, discretionary funding, and regional spread. Unlike the United Kingdom's system of university grants, the block funding of laboratories in France by the Centre Nationale de Recherche Scientifique, or the institutional support of the Max Planck Gesellschaft in the Federal Republic of Germany, the U.S. system would emphasize the support of specific projects. [14] Funds would be provided to the investigator or team to cover the direct costs of the project. The universities would be financially accountable.

Proposals would compete on their merits and be judged by those knowledgeable in the particular field. (The judges came to be known as *peers* and the system as *peer review* until NSF Director Erich Bloch dropped the term in favor of *merit review* in 1986.) The best science would thereby be supported and the national interest most effectively promoted. Since, however, bright people were located in universities all over the country, funds would be distributed among different regions. This would be done without the disadvantages of a state-by-state allocation formula.

The Republican sweep of the 1946 congressional elections moved the debate away from Senator Kilgore's conception and made the merit system possible. Kilgore was divested of his chairmanship of the Subcommittee on Science Legislation of the Military Affairs Committee and replaced by H. Alexander Smith, who was sympathetic to points of view expressed by friends on the Princeton faculty. They stressed the importance of quality in scientific effort and the need to insulate research as far as practicable from politics.

The system based on merit review and on individual projects has never been the whole of the research support system. It has coexisted with elements of institutional support, formula funds, and continued assistance from corporate philanthropy. Many large centers are in universities whose support has amounted to block funding. Large laboratories outside universities but administered under contract with university systems have provided management fees constituting institutional support. Projects, furthermore, can be small or large, and a large enough project may attract so much funding that it seems almost a form of institutional support. Even the National Science Foundation, which became the home of the small investigator-initiated grant reviewed by scientific peers, administered large programs at major facilities almost from its inception. Fellowship programs that evolved in the 1950s combined merit and institutional support (for example, the awards were portable and would pay tuition plus administrative fees to the host university).

Some institutional support to universities was also provided by other government agencies. Examples of support for centers or institutes include the Joint Services Electronics program, the National Magnet Laboratory, materials research laboratories, the Joint Laboratory for Astrophysics, the Harvard-Smithsonian Astrophysics Center, and the MIT Energy Laboratory. The Office of Naval Research also awarded contracts to Woods Hole, Scripps, and other oceanographic institutions. Project grants to principal investigators probably never constituted more than half the total federal funds given to universities in the heyday of research support in the mid-1960s. No precise determination of the point is possible, but a 1965 estimate of the approximate distribution of university research support gave 68 percent to project grants, 10 percent to insti-

Table 3-2. **Federal Research Support, by Major Category, Actual and Recommended Expenditures, 1963**
Percent

Category	Actual[a]	Recommended	Difference
Direct R&D support (project and coherent area grants)	68	50	18
Institutional support	10	25	−15
Training (pre- and postgraduate)	16	25	−9
Construction	6[b]	0	. . .

Source: Harvey Brooks, "Future Needs for the Support of Basic Research," in *Basic Research and National Goals* (Washington: National Academy of Sciences, 1965), p. 103.
a. Calculated as of 1963
b. Further federal construction was not needed at that time; operation and maintenance of facilities was provided under other grant categories, as part of indirect costs, or through nonfederal means.

tutional programs, 16 percent to training, and 6 percent to construction (table 3-2).

While the project system was never the whole of the federal effort, it determined the essential character of the postwar system. No class of institutions, including private and public research universities, had any entitlement to federal funds; rather, the government supported individual projects deemed meritorious in a competitive process. Some form of objective assessment by scientific peers was undertaken to assist in making decisions, whether the matter was funding for a large or small grant or contract, a research center, or a large facility.

While the agencies that funded outside research after the war did so with implicit attention to scientific merit, some (particularly the military services) relied on their own technical staffs for evaluations rather than an elaborate network of outside observers. The Office of Naval Research generally followed an informal review process, often just by phone call. There was never a standard peer review procedure followed by all grant-making agencies. As long as growth was rapid, few complained.

Clearly, the health of the institutions in which the research was to be conducted would have an important bearing on the long-term national capacity in science and technology. In its annex to the Bush report, the Bowman Committee, almost forecasting the controversies over indirect costs that erupted several decades later, warned of the difficulties of separating the costs of research and of instruction. It used this warning as an argument to support the concept of matching grants. But as long as the system grew rapidly such difficulties seemed minor. If an individual grant or contract were to shortchange the university, the total project support (plus fellowships and research assistantships) worked to the university's advantage.

There remained the issue of what the role of the mission agencies would be vis-à-vis that of the lead science agency. Bush and Kilgore, though disagreeing on whether the proposed NSF should principally support university research or applied research in the government's own laboratories, agreed on the important central role for the foundation in shaping the nation's research efforts. Bush did not want an independent medical research empire, preferring that medical research become a division within the foundation. And he wanted all civilian research on new weapons and military systems to be administered through a military grants division of the foundation (applied research on existing weapons would continue to be performed by the military services). The wartime experience had convinced him that military bureaucracies were inhospitable to creative research. Without the suspension of the normal procurement rules during the war, he thought, the Office of Scientific Research and Development could not have functioned. More broadly, he thought that basic research could not be administered effectively or performed within an agency that had operational responsibilities: "Operating agencies have immediate operating goals and are under constant pressure to produce in a tangible way, for that is the test of their value. None of these conditions is favorable to basic research. . . . Research will always suffer when put in competition with operations." [15]

It is a matter of mild irony that while the Kilgore and Bush forces battled over the structure of the NSF, events moved the nation's research system along a path different from what either side wanted. The resulting system emphasized the primary role of the large, mission-oriented agencies in supporting university research (and also applied research and development in industry). While the NSF was still being debated, the Office of Naval Research, created by statute on August 3, 1946, took the lead in funding basic research in the universities on a wide variety of subjects related to naval missions. [16] The ONR had an unexpended $40 million appropriation left from wartime construction funds as an initial budget, a strong staff built up for its wartime predecessor office, and excellent leadership. Using a simple contract mechanism, it quickly moved to support investigators in universities across the country and to institutionalize in the navy a program of support for basic research. The other services followed suit and established similar offices, most of whose funds were also devoted to basic research in the universities. [17]

The Atomic Energy Commission, also created in 1946, supported basic research as well. In particular, it administered programs in high-energy physics and later in fusion energy at large facilities. Most of the AEC budget, as with those of other mission agencies, funded development and production, and the commission operated or supported an extensive network of national laboratories (some affiliated with universities or university consortia, some not). But active support of basic research was judged a desirable complement to the more developmentally oriented technical efforts. Moreover, it was considered impractical to transfer the support for basic research in universities undertaken by the AEC, the National Institutes of Health, and the Defense Department and to disrupt the ties that had developed. Bush's opinion that research would be neglected if it were conducted in an operating agency was thus replaced with the idea that useful cross-fertilization can occur between development and basic research, provided that basic research had some measure of administrative independence from development activities.

Basic research broadly relevant to an agency's responsibilities came to be known as "mission-oriented basic research." [18] The term was partly an invention to ward off the criticism that the benefits of basic research were too remote to justify large-scale support and partly a reflection of the difficulty of describing in simple terms the complex links between discovery and application. The mission-oriented nature of the research meant that the system as a whole would not suffer from the disadvantages of rigid boundaries between different categories of research. Moreover, the system was ultimately driven by the needs of the users and not the producers of knowledge. Basic research would be generously supported, but the whole enterprise would be tilted toward the practical ends of government.

The mission agencies provided most of the initial federal support for basic research. In 1956 the government obligated an estimated $200 million for it. Of this amount somewhat less than $120 million went for basic research linked to national defense—$72 million to the Department of Defense, and $45 million to the Atomic Energy Commission. [19] If the National Advisory Committee on Aeronautics (after 1958 the National Aeronautics and Space Administration) were considered part of the defense effort, the dominance of defense agencies would be even greater. The National Institutes of Health reported $26 million obligated for basic research in 1956. The NIH, a mission agency in the sense that it was charged with public health responsibilities, has been distinctive in having its own large basic research and clinical medicine operations as well as providing significant support for basic research and clinical research conducted elsewhere. The division of funding between basic and clinical research has probably been about the same for the inside and the outside programs for most of the postwar period, about one-third for basic and the rest for clinical.

The initial reliance on defense agencies was evident also in the National Defense Education Act of 1958, passed in the wake of the Soviet Sputnik launches, that provided funds for university education and fellowships. The needs of national security, broadly conceived, were thus the focus of science policy before 1965. The high point in this dominance came in 1963, just before the buildup of the civilian space program, when 93 percent of all federal R&D funds came from DOD, AEC, and NASA. Most AEC support was divided among the weapons program, the submarine reactor program, and portable reactors for the army, and a nuclear rocket program for NASA. Much NASA R&D was joint military-civilian.[20] In the debate on science policy after the Reagan administration took office in 1981, some observers lamented the increasing share of DOD funds in total R&D, but this attitude reflects a short memory. The increase in defense R&D that began in the final year of the Carter administration and accelerated under Reagan was not a departure from tradition but a partial return to pre–Vietnam War practices.

The emphasis on national security inevitably meant that the Defense Department would be critically important to all aspects of science policy. But what was to be the role of the National Science Foundation, which the Bush report envisioned as the lead science agency, if even basic research was supported primarily by mission agencies?

An intriguing sidelight is to wonder what might have happened if the new foundation had not been vetoed by President Truman in 1947. The NSF almost surely would have had one of the wartime scientific leaders as its director, and there would have been a board of equal distinction. The funding patterns of the mission agencies and the division of responsibilities between NSF and the rest of the government could have evolved differently. The NSF might have expanded its function more rapidly and approximated the coordinating and leadership role envisioned by Bush and Kilgore. Whether it would have pursued the goal of promoting the rapid progress of science, as Bush and other scientists wanted, or moved toward accommodating the goals of society as set forth by nonscientists, Kilgore's preference, is a further question.

An understanding of how the system evolved must start from the recognition that science policy shared the pluralist character of American political institutions, and no strong central voice emerged. As Hugh

Heclo has noted, the United States is the only major nation with a "vacuum at the center" of its government.[21] There are no strong staff units serving the president, and the structure encourages most senior career executives to move from the center of policymaking toward the departmental periphery. Pluralism, existing as condition, is extolled as doctrine. The presence of many agencies to which an investigator can turn for support is deemed desirable because the situation protects the freedom of the investigator (who can go elsewhere if turned down by one agency), stimulates healthy competition and dynamism in the system, and prevents the rise of any autocratic centers of power. The role of the NSF must be seen in this context. Although the resources at its disposal at first were not large (they did increase tenfold from 1950 to 1960, starting from a very small base), the foundation was nonetheless important in funding general purpose basic research and in serving as a vital balance wheel among the various research support agencies.[22] The NSF has been responsible for the health of basic science enterprise as a whole, for superimposing a broad national perspective on the practices of individual agencies, and for filling gaps in the funding that resulted from the focus on mission-oriented research so that the quality of the nation's research and its long-term growth can be maintained.

This was a considerably weaker role for the NSF than many wartime leaders had envisioned, and they chafed at the limits, particularly the modest share of funding for basic research in the total national R&D budget (7 percent in 1960 and never, even allowing for idiosyncrasies of definition, more than 10 percent). There was a residual fear that basic research could be easily neglected if budgets were to become less generous, a concern reminiscent of Hoover's appeals for a national research fund and Bush's concerns that operational responsibilities would drive out research. In the mid-1960s, the NSF accounted for less than 2 percent of the federal R&D budget and for slightly less than 15 percent of all support for basic research.[23]

Some thoughtful observers believed the NSF support of core scientific fields would have to be more prominent. If it were truly to be the lead science agency and, in effect, the voice of American scientists, the foundation would need the money to set the tone and direction of the nation's scientific effort. These lingering concerns, however, were largely overlooked in the general mood of expansion. No serious challenge emerged to the view, stated in President Eisenhower's March 17, 1954, executive order, that the mission agencies should support basic research "in areas which are closely related to their missions" and that they should take a broad view of what research might be relevant. Later, NSF responsibilities would be broadened and those of the mission agencies narrowed. For the moment the consensus on the importance of basic research extended to reasonable agreement on how and by which agencies the research should be funded. Support of basic research by the agencies was a technical overhead on their missions, and the NSF budget was an overhead on the overhead.

Finally, the Bush report had recommended creation of a permanent science advisory board because of the need for "some measure of coordination of the common scientific activities of [federal] agencies." The board was to "advise the executive and legislative branches of the Government as to the policies and budgets of Government agencies engaged in scientific research." The report further suggested that the board be composed of "distinguished scientists who have no connection with the affairs of any Government agency."[24] The NSF bill, passed by the Congress in July 1947 and vetoed by President Truman, contained a provision for an interdepartmental science committee made up of representatives from the leading science agencies and chaired by the director of the foundation. The attempt to coordinate the budgets and policies of federal agencies through an official not responsible to the president was patently unworkable and was, indeed, one of the major reasons for the veto. The Steelman report of August 1947 refined the idea to include an interdepartmental committee created by executive order and reporting directly to the president, a review unit in the Bureau of the Budget, a science liaison official on the White House staff, and a politically accountable NSF.[25]

Evolution of the Government
Research Establishment

In addition to federal funding of basic research, the postwar consensus also covered the responsbilities of the government research bureaus. In general, it was decided that the technical agencies would have more duties but would expand along traditional lines. The universities would do the lion's share of the most important basic research and would set the direction of the nation's scientific efforts. Industry would receive the bulk of federal funds for development—the largest part of the R&D budget—and would supply most of the technological systems produced for government use (some few items, such as heavy cannon and certain categories of ammunition would continue to be produced in government facilities). Support for extramural R&D would be an important impetus to the emerging contract state.[26] These activities would reinforce and complement the agencies' in-house technical efforts. To orchestrate the system, the government's own technical resources and expertise would have to be of a higher order than before.

The government's in-house efforts would continue, as before the war, to center on applied research relevant to agency missions. But the research would be on a larger scale, more fully articulated institutionally, and more integrally related to agency goals. Program managers, scientists, and contract administrators within the government would all have added responsibilities. The military services perhaps illustrate the point most clearly.

The army and navy laboratories and research programs were greatly expanded during the war. But unlike the post–Civil War demobilization and the partial dismantling of technical activities following World War I, in 1945 it was obvious to military planners that a permanent revolution had occurred in the relationship of technology to warfare, and continous mobilization of scientific and technical resources became a national strategy.[27] The military now had the responsibility to nurture mission-oriented basic research in the universities.[28] In addition to their extramural research support, the services themselves conducted some basic research to strengthen their technical capability and teach themselves to be discerning users of contract research. In fiscal year 1986, for example, the Naval Research Laboratory spent some $500 million on basic research, half for extramural projects, mostly in universities, and half to support in-house research. The budget was far lower just after World War II, of course, but the proportion has remained remarkably stable. And the share of government's expenditures on research and development conducted by federal agencies themselves has also remained stable, about one-quarter of the total budget. Table 3-3 shows expenditures for federal R&D by performer and as a percentage of total federal R&D expenditures.

The services also support exploratory and advanced development, demonstrations of technical feasibility, proof of concept, and early design of systems and the like through contracts with outside performers. As work proceeds toward development and eventually to production and deployment of systems, industry gets more contracts than universities. In-house laboratories continue to complement the efforts of contractors at every stage.

Funding for basic research and for exploratory development may be

Table 3-3. **Federally funded R&D, by Type of Performer, Selected Years, 1967–86**
Billions of dollars and as percent of total

Type of performer	1967		1971		1976		1986	
	Billions	%	Billions	%	Billions	%	Billions	%
Federal agency intramural	3.4	23	4.2	28	5.6	28	13.5	24
Industry	8.4	58	7.7	52	10.2	51	30.0	52
Universities	1.4	10	1.7	11	2.5	12	8.1	14
Federally funded R&D centers	0.7	5	0.7	5	1.1	5	3.6	6
Other nonprofit institutions	0.6	4	0.7	5	0.8	4	2.1	4
Total	14.5	100	15.0	100	20.2	100	57.3	100

Source: National Science Foundation, *National Patterns of R&D Resources, Funds, and Manpower in the U.S., 1953–1976* (Washington, 1987), table B-1; and National Science Board, *Science and Engineering Indicators, 1987* (Washington: National Science Foundation, 1987), tables 4-1 and 4-4.

combined administratively, as in the navy where the chief of naval research handles both functions through the Naval Research Laboratory, or they may remain separate, which is the pattern of the army and the air force. In either case, there is a problem of effective links between the categories and of knowing when a transition should be made between categories. Because the categories overlap, much of the manager's job is to identify when a concept is ripe for further development or what kinds of fundamental inquiry can usefully support advanced development projects. This is the government's analogue to the interfaces between research, product development, production, and marketing that confront corporation managers.

Critical aspects of R&D require in-house efforts rather than contractors. Testing and evaluation, including field tests under simulated operational conditions, are conducted by specialized units within the services, usually separated from, though working closely with, the operating commands. Contractors with special competence in testing and technical evaluation may be involved, but the services guard their prerogatives and tend to rely on their own staffs in evaluations because the work includes reviews of contractors' performance. When a weapons system tests successfully, the production stage begins. Production normally involves far larger expenditures and new technical and administrative complexities as the service chooses a contractor to produce defense systems. Production may involve some overlap and repetition of earlier stages, which may have been only preliminary.[29]

The missions of the services have been supplemented by the increased technical capacity of the Office of Secretary of Defense. The Advanced Research Projects Agency (now the Defense Advanced Research Projects Agency, or DARPA) is concerned both with technology development at the interfaces between the services' common technologies and with supporting scientific fields important to long-term military objectives that may have been neglected by the service funding authorities. DARPA came into being early in 1958 as part of America's response to the Soviet space challenge. It was functioning six months before NASA was established, and under its chief scientist, Herbert York, decided the major directions of the military space effort and began the planning for civilian projects that it later transferred to NASA.[30]

Recognizing the impact of technology on military missions, the individual services as well as the Office of the Secretary of Defense have also established organizations that perform analytic and planning functions. The first of these, and still the best known, is the Rand Corporation, created in 1946 to perform long-range research for the air force.[31] Rand's initial study was of the military implications of an earth satellite, but later studies increasingly reflected broader assessments of the strategic dimensions of technological developments. Similar groups have provided contract analytic services to the army, navy, and Office of the Secretary of Defense.

The Defense Department thus has a galaxy of staff units, laboratories, contract research centers, extramural funding authorities, and other entities engaged in applying technology to modern warfare. Recognized career paths exist for officers within the technical operations of the military services (though the most senior positions are still held by those with extensive operational command experience).

Though the most sophisticated consumer of technology, the military is by no means unique. Despite variations in nomenclature and function, many other agencies have developed highly articulated systems for supporting research. The arrangements may resemble the military's. A given agency may fund some mission-oriented basic research in the universities, perform some in its own laboratories, engage in applied research and the early stages of development, and monitor the industrial contractors who perform most engineering or production work. NASA is an instance of a pure technology agency whose mission is to advance particular technologies and achieve technological goals.

There are, of course, differences among agencies. Those created before the war, when virtually all government-sponsored research was in-house, tend to fund much less outside research. Even during the Apollo buildup, NASA (an outgrowth from the old National Advisory Committee on Aeronautics) supported a much smaller proportion of university research than did DOD. The air force, declared a separate service after the war, has supported much more outside R&D than the army has. NSF has no laboratories of its own and no complex procurement activities with industry, although it does support a number of federally funded research and development centers that resemble in-house laboratories. NIH provides grants and contracts for research activities but does not procure anything other than routine supplies. The technical functions of the Department of Commerce are an elaboration of its historical missions—awarding patents, defining standards, predicting weather, and disseminating technical information. The Department of Energy (the Atomic Energy Commission just after the war and later the Energy Research and Development Administration) operates an extensive network of national laboratories, funds and manages the nation's high-energy physics program, and at various times has sought to carry development as far as the demonstration project. The department and its predecessor agencies have provided technological services for national defense (notably in producing nuclear weapons), biomedical scientists, electrical utilities, and consumers. The Environmental Protection Agency is principally interested in scientific activities related to its regulatory, risk-assessment, and substance-testing functions.

World War II ushered in broader government technical activities and improved professionalism of its scientific staff. Agencies' R&D functions became a permanent part of their activities and a significant part of an expanded national research effort, overlapping to some degree the functions of other institutions but in general following the path sketchily mapped in the Bush report. Bush had stated that, with some notable exceptions, "most research conducted within federal laboratories is of an applied character. This has always been true and is likely to remain so." Nonetheless, "research within the Government represents an important part of our total research activity and needs to be strengthened and expanded after the war." [32] The research was to focus on matters of special public importance, particularly those on which the private sector was doing little, either because industry foresaw limited potential for profit or because universities found the problems of insufficient theoretical interest.

The role of government research has been remarkably stable. Although organizational changes have been frequent, research priorities have shifted, and the fortunes of individual organizations have waxed and waned, the system has displayed continuity. Government laboratories have been the object of periodic management scrutiny but have usually emerged intact and with their missions only slightly modified.[33] Agency responsibilities for the support of certain research areas have also changed. The NSF assumed the lead in funding basic research in oceanography after the navy stopped its logistic support of the antarctic research program. The foundation has also become the leading supporter of research in atmospheric sciences and molecular biology and biochemistry, although the navy, the National Oceanic and Atmospheric Administration, and NIH continue to have strong programs. Funding for the large-scale radio astronomy facilities has also shifted to NSF. Certain engineering fields, once heavily dependent on the Defense Department, have come to depend on NSF as well, although during the Reagan administration, Defense support of university engineering recovered, especially in artificial intelligence, robotics, and computer-aided manufacturing. NIH has steadily become a major source of funding for academic research in the life sciences. Such changes have not, however, greatly affected the missions of the government research laboratories, which continue to support the goals of their agencies even as new policies direct emphasis from one field to another or as funds available for extramural research vary.

The role of the government research establishment in carrying out science policy has thus largely followed the pattern agreed upon in the postwar consensus. The nation set out to enlarge and strengthen the technical activities of all agencies whose missions were strongly affected by developments in science and technology. At first, this meant national defense and related agencies in particular, but social needs would eventually include most others, for advances in technology had a pervasive impact on traditional missions and jurisdictional boundaries. Applied research was emphasized, but government scientists drew on the best basic research and sought to keep up with the most advanced industrial applications.

These purposes and goals have largely been accomplished, and the policy has not required fundamental modification. While other elements of the postwar consensus have been subject to intense debate, the role of the government research establishment has been less controversial. Its two most controversial aspects emerged during the debate in the 1960s over contracting for R&D with the private sector, and in the 1980s over the role of national laboratories in technology transfer. In the contracting dispute, critics contended that the government had unwisely delegated to private think tanks and federally funded research centers the critical management responsibilities that properly belonged with accountable officials. The controversy triggered an inquiry under the chairmanship of David E. Bell, director of the Bureau of the Budget, that resulted in a multivolume report largely defending existing practices but urging that the government strengthen its management capabilities.[34] The controversy over technology transfer in the 1980s will be left for later consideration.

Questions will continue to arise about the morale, quality, and efficiency of the government's scientific work force and on how well research units perform their assigned missions. But the place in the national system of the government's research establishment, firmly rooted in tradition and evolving along established lines, is secure.

Commercialization of Technical Discoveries

The postwar consensus also entailed agreement on the way technology moves from laboratory to commercial application. The central idea was the absence of an idea: there seemed to be no need for a self-conscious strategy to promote innovation. Doubtless many things could affect the pace and diffusion of technical discoveries, and since policy decisions could affect the framework for discovery, the matter should not be viewed in absolute terms. Fundamentally, however, the consensus saw little need for explicit policies to foster the development of nonmilitary technologies as long as government encouraged market incentives. If American science flourished and personnel shortages were corrected, then commercial applications based on technical innovation would occur more or less automatically.

The process began with the universities, which would train scientists as a by-product of conducting basic research. Some of these people would join industry and spend their working lives in innovative activity. The government's own research activities would stimulate the private sector when necessary by focusing attention on national priorities, demonstrating the feasibility of advanced technology and in some cases providing a market (at least a start-up market) for goods produced by industry. Providing a start-up market was perhaps most evident in government encouragement of aircraft development and in the early phases of computer manufacturing.[35] The aim here was to assist in developing products and in particular to sustain the technology useful to the national security agencies. Creating commercial products and markets was incidental. Moreover, the government functioned mainly as a buyer, leaving wide latitude to industry in the actual design of the products.[36]

Beyond making these assumptions, neither the Bush report nor the Steelman report, nor President Eisenhower's Science Advisory Committee devoted significant attention to the problem of industrial innovation. Bush's opening sentence in the section dealing with industrial research stated, "The simplest and most effective way in which the Government can strengthen industrial research is to support basic research and to develop scientific talent."[37] The Steelman report added that the nation was certain to confront "competition from other national economies of a sort we have not hitherto had to meet." Wartime destruction, it reasoned, would compel other nations to rebuild their industrial plants with the most modern technologies, and many would also pursue state-directd strategies. The answer to the challenge was progress in the basic sciences: "Only through research and more research can we provide the basis for an expanding economy, and continued high levels of employment."[38] President Eisenhower's Science Advisory Committee reiterated that flourishing basic research and strong graduate education, together with the normal incentives of the marketplace, would protect the nation's technological leadership.[39] Not until the Kennedy and Johnson administrations did these beliefs begin to be challenged and then only inconsistently and uncertainly, leaving the fundamental critique to the Nixon administration.

Other science matters recognized as important to commercial development were tax and patent policy. On these points, however, there was no clear agreement. Bush pointed to the income tax law as "one of the most important factors affecting the amount of industrial research" and observed that "uncertainties as to the attitude of the Bureau of Internal Revenue regarding the deduction of research and development expenses are a deterrent to research expenditure." He recommended revision of the Internal Revenue Code "to remove present uncertainties in regard to the deductibility of research and development expenditures as current charges against net income."[40] Industry's desire for immediate deductions for purchases of research equipment collided with broader tax policy objectives, however, and it did not win its case. Differences between industry and the Internal Revenue Service over the proper tax treatment of R&D equipment continued until the 1980s.[41] Some analysts also regard the tax treatment of executive compensation, including stock options and capital gains, as a critical factor affecting entrepreneurship. Management's willingness to invest in new technologies, they argue, is affected by the differential between the tax rates for ordinary income and for capital gains.[42]

Bush's views on patent rights reflected the considerable uncertainty left over from Senator Kilgore's wartime efforts to force large companies to disclose technological data gained under government contracts. Bush, himself a strong believer in small business, sympathized with many of Kilgore's ideas, but he realized that in its original extreme form the senator's proposal could undermine intellectual property rights and discourage technological progress. Bush had, in any event, to fight against the massive disruption of the war effort the Kilgore proposal might entail. But once the war was over, what would be the best long-term policy? Bush's views here were a muddle. In *Science, The Endless Frontier,* he seemed to suggest that because of "uncertainties" in its operations the patent system hurt the nation by impairing "the ability of small industries to translate new ideas into processes and products of value to the nation." He went on, however, to suggest that the system was "basically sound" and only "certain abuses" had led to "extravagantly critical attacks" designed to discredit it.[43] Finally, Bush took refuge in a Commerce Department review of the issue and urged that action on the patent laws be withheld pending completion of the detailed study.

Not surprisingly, patent rights remained a refractory issue for many years. Agencies pursued different patent policies as they expanded their research efforts. The Defense Department became the representative of the liberal policy, allowing contractors to purchase royalty-free the rights to their discoveries and to apply for patent protection through normal channels. (The process of granting patents, as such, has not been a subject of controversy.) The Atomic Energy Commission stood at the opposite pole, insisting on government retention of patent rights to all discoveries made under an AEC research contract. The logic for the restrictive policies, which were imposed by legislation, was never clear but reflected a concern with secrecy and with regulatory objectives and a carryover of Kilgore's populist view that the government should own all rights to the research it pays for. Most agencies with outside research programs fell between the positions of the AEC and DOD. Despite high-level efforts to harmonize agency policies, including a 1949 review by the attorney general, a 1963 statement by President Kennedy, and a 1972 statement by

President Nixon, in addition to periodic reports and studies at lower administrative levels, no uniform policy emerged.[44]

Not until the passage of the Patent and Trademarks Amendments in 1980 did the situation improve. The amendments consolidated twenty-six policies, executive orders, and statutory guidelines covering patent rights for those performing R&D for the government into a coherent framework providing for the easier exploitation and ownership of technical discoveries. The Patent Act and its amendments extended the principles of the Institutional Patent Agreement to all government-sponsored research, enabling universities freely to patent and license discoveries made in their laboratories with the aid of public funds. Thus the mingling of public and private funds in the same project was no longer a problem. The universities were also allowed to grant exclusive licenses. The Small Business Act (part of the amendments) also directed broader changes in the relationship of universities to small businesses. Federal R&D agencies were required to designate up to 1.25 percent of their grant funds for research to be performed by small businesses, and up to 30 percent of those funds were to be directed to joint research ventures of universities and small businesses. The universities initially fought the proposal, but they have benefited substantially from the relationships stimulated by the act.

Since the 1980 legislation the thinking has been that the inventing organization is the one most likely to exploit the opportunities arising from its research. Given the importance of strengthening American industrial competitiveness, it must be encouraged to do so. Patent policy should therefore promote American entrepreneurship. The more intense concern with intellectual property rights in recent U.S. trade policy also evolves from these patent issues.

Finally, the reports of the 1940s and 1950s assumed the government would pursue "correct" macroeconomic policies, which meant that high aggregate demand would promote a favorable investment climate, economic growth, and full employment, as set forth in the Employment Act of 1946.[45] Exchange rate swings were not a problem; under the Bretton Woods regime, fixed par values for currencies, which were pegged to the dollar and to gold, would ensure stability in international transactions. The international economy was to be an outlet for U.S. goods when domestic demand slackened. Inflation was a residual concern, too, especially when wartime controls were lifted. But aside from the immediate reconversion period after World War II and again after the Korean War, inflation was not a threat.

Industry, in short, was able to assume a stable environment, with domestic demand steadily rising for an endless stream of new products. The
international marketplace was also emerging as part of the endless frontier of economic opportunity. The general confidence in America's ability
to manage macroeconomic policy and to ensure a stable environment for
investment and economic growth was high and remained so until the
Vietnam War overheated the economy in the 1960s.[46] As the war intensified, inflation increased, and the monetary disturbances that were to
undermine the Bretton Woods system grew in severity. Macroeconomic
policies could no longer ensure a favorable climate for technological innovation simply and with certainty. The influences of macroeconomic
and microeconomic policies on innovation and competitiveness have
now become among the most refractory issues for the analyst and the
policymaker. The optimism that underlay the Bretton Woods system
could not be sustained. As the OECD economies recovered, their productivity growth rates began to catch those of the United States, and the
comfortable image of American industry as undisputed world leader in
advanced technologies was shattered. Beginning in the 1970s, trade issues assumed more importance in policy debate, with mounting pressure
on U.S. administrations to protect American industry from unfair trade
practices.

Science and Regulation

The fourth element in the nation's postwar science policy was shared
attitudes toward the regulation of scientific and technological activity.
Like government support for commercial innovation, regulation was not
generally viewed as an urgent matter for public policy to address. Science
and technology were leading people toward a fuller life: disease, poverty,
and other traditional scourges would be conquered. The cure for the ills
of science was more science. Of course, the founders of postwar science
policy recognized that some wartime innovations would need to be controlled, but this was of less importance than the policies to promote
science.

To understand how regulation was viewed, three dimensions of the
issue must be distinguished. First, traditional grounds for regulation assumed that new technology required creating appropriate government
mechanisms to monitor, regulate, or oversee it to protect health, safety,
and the environment. This had happened with the introduction of the
steamboat and railroad, the expansion of the food processing industry in

the early part of the twentieth century, controlling air traffic in the late 1920s, and overseeing the pharmaceutical industry in the late 1930s. The second dimension was the effectiveness of the methods and processes of science employed to detect and measure the risks that might arise when new technologies, substances, or materials were introduced into the workplace or the environment. Third, how much regulation would scientific research itself require? Research might involve the release of harmful organisms into the atmosphere, unethical experimentation on humans or animals, systematic denial of career opportunities to women or minorities, or misappropriation of public funds. More broadly, how much regulation did the institutions in which science was conducted—notably the research universities—require in the name of accountability or efficiency?

With respect to the first dimension, the nuclear revolution of World War II obviously called for a regulatory response of some kind. The reaction to issues raised by the creation of the atomic bomb typified the attitude of the public just after the war. The underlying assumptions were that technology, fueled by research, was rushing forward. Most developments were to be welcomed as contributing to human betterment. If the technology were important to public objectives, government policy might assist development. A major new technology, such as that stemming from the release of atomic energy, would also require regulation: obviously the spread of nuclear technology would need to be controlled to ensure national security.

The Acheson-Lilienthal plan embodied such principles of control, including America's renunciation, under appropriate safeguards, of nuclear weapons. Control was also sought by establishing the Atomic Energy Commission, created by the Atomic Energy Act (MacMahan Act) in 1946, along with a joint congressional committee on atomic energy, to regulate (and promote) peaceful uses of nuclear energy.[47] There seemed nothing odd in having the Atomic Energy Commission promote the use of civilian atomic energy and at the same time work to control harmful side effects: it had the expertise to do both. After the collapse of the Acheson-Lilienthal initiative, a group of members of the Federation of American Scientists, which had backed the plan, advocated a moratorium on the development of civilian uses for nuclear power on the grounds that material for making bombs would become too accessible. They were decidedly a minority (though their concerns were revived during the Carter administration in connection with the development of fast breeder reactors and the reprocessing of nuclear fuels).

Because instances of technology as a danger rather than an opportunity seemed rare, however, there was no point in imposing unnecessary regulatory burdens on the economy. Objections to the merger of promotional and regulatory functions within the same agency, inadequate safety standards, and failure to provide adequately for nuclear waste did not arise until much later. Whether nuclear fuels, moreover, were given unfair precedence over research on other fuels because AEC research funds focused heavily on nuclear matters was also an issue for the future. For the moment, regulating a new technology and promoting its development were not regarded as incompatible objectives.

Further, the dangers that might be created by new technology were considered in terms of acute episodes rather than chronic risks. Those who worked with atomic energy, of course, had to be protected against accidents or excessive exposure to radiation. The industrial pollution in Donora, Pennsylvania, in October 1948 that caused 20 deaths and 6,000 illnesses, and the choking smog that increasingly afflicted Los Angeles under certain atmospheric conditions also demanded attention. But the answer lay in taking immediate public health measures and conducting long-term research to understand the causes of the problem.

The Delaney clause of the 1958 Food, Drug and Cosmetic Act, which regulated food additives, is a classic instance of this approach to regulation. The clause, part of three amendments adopted to define the regulatory system for the Food and Drug Administration as it implemented the 1938 Food and Drug Act, established a system whereby the FDA could ban or limit the use of a specific substance in or on food. The Pesticide Chemical Residues Amendment of 1954, now section 408 of the Food, Drug and Cosmetic Act, provided that a raw agricultural commodity shall be deemed adulterated if it bears any residue of a pesticide that does not conform to an established tolerance. The Food Additives Amendment of 1958, now section 409, established a licensure scheme for substances intended as ingredients in formulated foods (and for packaging material) and specified that the agency may permit the use of food additives that are "safe," a standard described in the legislative history as requiring reasonable certainty that no consumer of a food containing the additive will suffer harm. The Delaney clause prohibited the FDA from declaring as safe any food additive found to induce cancer in humans or in laboratory animals.[48]

While the standard thus established was redundant—other provisions banned carcinogenic substances—the Delaney clause came to typify the absolutist approach to regulation in the 1950s. When a substance was found to produce cancer, it was banned. The assumption behind the legislation was that although some few substances would inevitably be too dangerous, it was a relatively straightforward matter to identify and ban them outright. The far more pervasive impact of long-term risks to health and the environment from carcinogens and other toxic substances, and the complexities of risk-benefit analysis of the more numerous substances posing some detectable level of risk, came later to the public debate. By holding on to ideas and practices from a more optimistic past, policymakers postponed facing the inevitability of hazards created by technology.

The methodologies involved in the regulatory process reflected other assumptions. There was less of the sense of urgency that would later give rise to such applied fields as risk assessment and other interdisciplinary efforts to assist regulators in solving immediate problems. Instead, the focus was more on basic research, which would eventually clarify troublesome issues and make effective regulation possible. Government laboratories that engaged in research on air pollution after the 1955 Clean Air Amendments, for example, focused mostly on fundamental research.[49] To be effective, it was assumed, regulation had to come later and must be built carefully upon the best scientific knowledge. Similarly, the extent to which research itself needed regulation was not an urgent issue. There was little fear of the mad scientist threatening the lives of millions when scientists convened in 1981 at Asilomar, near Monterey, California, to announce a moratorium on research into techniques of recombinant DNA.

International Context of Science Policy

Attitudes toward science and technology also had implications for the conduct of foreign affairs. For example, since basic research was assumed to be important, the federal government would have a responsibility to support, at least on a modest scale, basic research overseas. Further, sometimes the logic of a scientific field dictated international cooperation. Radio astronomy required observation points around the world. In the 1950s the NSF also began to support an optical and infrared observatory in Cerro Tololo, Chile. Cooperation was also necessary for earthquake observation in foreign nations and for oceanographic vessels to visit foreign territorial waters. The International Geophysical Year of 1958 was a spectacular success story of collaboration among nations to promote scientific advance.[50]

Often the motive behind cooperation was national prestige and the desire to conduct cultural diplomacy. American science was widely admired, and exchanges of data or of scientists would promote U.S. interests in the broadest sense. The United States had also been the beneficiary of the greatest emigration of scientific and technical talent in history— the brain drain of scientists fleeing from Nazi Germany in the 1930s and a second wave seeking the favorable working conditions in America after the war.[51] Indeed, American science had benefited greatly from contacts with the world's scientists from the early days of the Republic. Acting now as a magnet for talent represented both good will and broad self-interest.

The government's own research agencies also pursued activities abroad, usually on a small scale and clearly subordinated to domestic priorities. The Department of Agriculture sponsored exchanges of information on crop disease, the Geological Survey made cooperative arrangements with counterpart agencies in other nations to monitor earthquakes, NASA included foreign payloads on satellite launches, and the Weather Service exchanged meteorological data. Each department or bureau, in short, had its own small foreign service.

With respect to the commercialization of research, the postwar consensus saw little need to depart from relying on domestic market forces. It was assumed that the noncommunist nations would operate within the framework of a free international marketplace. Congress initially balked at the implied infringement of sovereignty when the International Trade Organization was proposed in 1946 as the mechanism to regulate trade. But the 1947 General Agreement on Tariffs and Trade (GATT) became a satisfactory alternative for breaking down trade barriers and promoting freer trade among member nations.[52] American industry generally endorsed the new framework because postwar reconstruction was providing it with new opportunities to sell American goods and know-how.

Imports were not a serious threat to American jobs or companies. The Steelman report warned that the United States would face serious competition as other countries rebuilt their economies with the most modern production technologies, but it rejected protectionism and counseled reliance on maintaining leadership in technology and producing superior products through strong research efforts. The GATT regime would enforce trade rules and allay the report's fears of government-assisted trade expansion. It would police the system and provide remedies for injuries resulting from restrictive practices. The Bretton Woods agreement would provide a stable currency framework for international transactions.

American companies were quick to seize overseas opportunities; by 1960, more than 2,000 manufacturing plants had been established abroad, principally in Western Europe.[53] American technology, like American science, was a dominant force, and America's industrial success reaffirmed the value of basic science. Together they guaranteed that the American system was the right formula for success in the scientific age. The conception of a high-technology society—and even specific aspects of the formula—impressed foreigners. For example, many nations attempted to spend 3 percent of GNP for research and development. They also tried to strengthen basic research and university systems, adopt project support for research, and allow wide latitude for independent investigators to choose their areas of inquiry.[54]

The technology gap between the United States and Europe, as diagnosed in the 1960s by Jean-Jacques Servan-Schreiber, resulted from Europe's neglect of science and technology in its educational system, its management shortcomings in contrast to America's ostensible capacity to plan and manage innovation, and cultural attitudes that blocked change.[55] The solution, he wrote, was to outdo the Americans at their own game; but for other Europeans, it did not appear possible simply to imitate the Americans. A smaller industrial nation could not achieve the breadth of coverage that the U.S. program for research support had. It could, however, pursue a more selective strategy, modifying elements of the American approach to fit its own traditions, or it could join in cooperative scientific ventures with other European nations.[56]

The process by which technology spread to the third world was explained in the theory of the product cycle, developed by Raymond Vernon of the Harvard Business School.[57] The theory posited that advanced products would be developed by companies with strong research efforts (many of which would, of course, be American). Those products would dominate markets until replaced by the next generation of goods embodying a more advanced technology. The earlier generation of products would then be marketed in third world nations via licensure or other appropriate mechanisms. The foreign company or government entering into the licensing arrangement would acquire the technical skills to manufacture the product. After a while these skills would diffuse through the local economy. As the process was repeated with each new generation of products, third world nations would gradually become more fully integrated into the world economy. The implication was, however, that they would be integrated as secondary markets for the goods of the industrial nations, not as serious competitors in the export of manufactured goods.

The theory of a product cycle was not entirely wrong in that the United States, drawing on a highly skilled labor force, has continued to produce many of the most advanced versions of new technologies. The theory, however, failed to consider the speed of technology diffusion, which has enabled Japan and some developing countries to capture mass markets for many of the new technologies pioneered by American firms, thereby undercutting the financial base for the R&D and investment necessary to develop the next generation of processes and products. Short product cycles, the rapid diffusion of process technologies, and intricate new patterns of global technological cooperation became the norm. The manufacturing expert visiting a production center for integrated circuits in Taiwan, Singapore, South Korea, Japan, or the United States can now detect few differences. But when the product cycle theory was posited, the international economy looked more favorable to U.S. interests and offered the almost limitless opportunity envisioned in *Science, The Endless Frontier.*

Economic expansion, moreover, whether in Europe or the third world, took place against the backdrop of American military power. It was not exactly a case of commerce following the flag: America supported decolonization and pressured European allies to accelerate their plans for it. But beginning with the signing of the Rio Pact of 1947 and the creation of NATO in 1949, America knitted together a system of regional alliances to contain Soviet power, maintained the freedom of the seas for commerce, and extended its nuclear umbrella over the entire "free world." Even the nonaligned nations, which resisted inclusion in any formal alliance, enjoyed the benefits of the relative stability afforded by the bipolar international order. A pax americana more or less prevailed and contributed to the flow of trade, investment, ideas, technology, capital, and people around the world. Indeed, in Robert Gilpin's analysis, a hegemonic nation was necessary after the war to ensure the orderly flow of goods and service transactions and to act as a market as well as lender of last resort.[58] Without such a force, the disequilibria in the system would always threaten to disrupt the flow of goods and disturb the workings of an international division of labor.

The cold war, however, limited the flow of technologies and the openness of the system. The Bush report had urged rapid demobilization and the lifting of wartime security restrictions because "there is no reason to believe that scientists of other nations will not in time rediscover everything we now know which is held in secrecy. A broad dissemination of scientific information upon which further advances can readily be made furnishes a sounder foundation for our national security than a policy of restriction which would impede our own progress although imposed in the hope the possible enemies would not catch up with us."[59] For a time, Congress seemed to agree with the need to ease wartime controls on most aspects of U.S. trade. But with the heightening of cold war tensions, export controls were extended. The Export Control Act of 1949, together with separate legislation dealing with atomic energy and munitions, imposed restrictions on the flow of technologies that could have military applications.[60] The Coordinating Committee on Multilateral Export Controls (CoCom) was also established at this time to administer the control arrangements with other nations.

The implementation of the controls did not seriously disrupt world trade. The fractious disputes with allies, the vociferous complaints of American firms against the controls, and the anxieties of defense analysts about technology leakage that have become common in recent years were largely absent. The reasons lie partly in the technological dominance of the United States. Its virtual monopoly of critical technologies permitted effective control, and allies so dependent on American economic and security assistance were not inclined to complain. That civil sector technologies continued to lag behind the most sophisticated military applications and thus did not require control was a further reason, as was the confident attitude that American technology would remain superior whatever minor leakage occurred. Finally, low levels of commercial activity between Western firms and Eastern bloc nations and the relative isolation of the Soviet bloc from the world economy simplified matters. As these factors changed, administering the controls became more contentious and a more important policy concern.

The relative calm of the first decade under the controls is illustrated in the startling Atoms for Peace program of the mid-1950s in which the United States attempted to spread nuclear knowledge, data, and technologies around the world despite the strict nuclear secrecy enshrined in law and national policy.[61] The purpose was to advance the peaceful uses of atomic energy, win potential markets for U.S. technologies, and gain prestige by demonstrating American technological prowess. The compartmentalization of thinking about military and civilian applications was such that policymakers saw no contradiction between peaceful use and the potential spread of militarily useful fissionable materials. The Eisenhower administration wanted to demonstrate U.S. interest in the civilian applications of nuclear energy to counter the image created by the U.S. policy of massive retaliation—the willingness to use nuclear weapons to contain communism, if necessary even in local conflicts. This political necessity overrode the risk—at that time considered minor—that the spread of generic nuclear technology would result in the proliferation of nuclear weapons. Few understood that reactor-grade plutonium could be used to make a bomb only slightly less powerful than those in the U.S. arsenal. When prevailing wisdom later decided there was no safe barrier between military and civilian nuclear technology, the regulatory objectives of policy became more pronounced. The desire to advance American prestige through technology also became a less important objective of foreign policy. The original Atoms for Peace goals gave way before those of the International Atomic Energy Agency, whose major function was to prevent the diversion of fissionable materials from civilian reactors to military purposes.

Science policy in international affairs is principally an extension of national concerns. Because for several decades after the war domestic policies were more promotional than regulatory, the same priorities prevailed in international dealings. When the U.S. environmental movement gathered momentum in the late 1960s, protecting the environment everywhere also became policy. Special complexities in enforcement, standards, and adjudication arise, of course, with international regulations. The timing and progression of environmental activism in various nations and conflicts over the extraterritorial reach of national law contribute to complexities of multilateral diplomacy.

Finally, matters such as satellite communication, remote sensing from space, international commercial aviation, oil transport supertankers, and frequency spectrum allocation are inherently of international concern. The commercial activity they generate inevitably involves government-to-government or enterprise-to-government relations that depart from free market conditions. While theories of the international context posited the same kind of free markets and open competition found at home, the analogy to domestic conditions was by no means exact. Inventors, technical managers, and entrepreneurs were assisted in important ways by their governments in coping with the vagaries of other governments. Multilateral agreements or regulatory regimes, painstakingly negotiated, have often been a prerequisite to commercial activity abroad.

Those parts of the domestic economy such as agriculture or telecommunications that were closely regulated at home would naturally be so in the international context. And if there were a free market at home but public ownership or regulation of a sector abroad, business had to adapt marketing strategies accordingly. These departures from a pure model of goods and capital flowing freely around the world seemed to American businesses operating overseas simply a matter of coming to grips with reality: with luck, perseverance, and the improved conditions around the world wrought by technical advance, America would prevail or at least prosper. At a deeper level, as Louis Hartz foresaw, the traditional grip of Lockean liberal values on American thinking blocked full understanding of the convulsive changes shaping the postwar world.[62] Foreign affairs and foreign policy were largely the province of an elite educated at a few Eastern private schools and universities. A more open policy process featuring the carryover of fractious domestic political disputes and new, and sometimes clamorous, voices would shortly take shape, but for the moment the postwar formula seemed to fit foreign as well as domestic policy.

Summary

The American research system took shape as the nation moved from demobilization to reconstruction of the world economy to stable prosperity, and from cold war tensions to the Korean War to protracted superpower rivalry. The elements all seemed in place. In 1960 the United States spent more than any other nation on R&D as a percentage of GNP, dominated Nobel Prize lists, and continued increasing its labor productivity and economic growth (despite cyclical downturns in 1954 and 1957). American goods and influence spread worldwide. The ideas that sustained the system were the kind most congenial to the American mind: they worked. Who could quarrel with success?

It was, in fact, a prodigious achievement. Instead of the Depression that followed the First World War, spectacular prosperity followed the Second. Businesses that planned for expansion flourished, and those pursuing a cautious strategy fell back.[63] The prosperity spread to the entire West, largely eliminating poverty for a quarter of the world's population. The more developed of the less industrialized nations and some new industrial countries shared in the growth. Living standards improved and life expectancies increased despite continued dire poverty, periodic famine, and disease in some developing nations. Science and the arts flourished. Political stability, progress, and hope for the future could lead one to conclude, "This would be America's century."[64]

Yet by 1970 the social, political, and economic landscape had changed dramatically. The consensus on foreign policy lay in wreckage. The assumptions of science policy became subject to intense scrutiny and doubt. Although cushioned somewhat from the shocks affecting the political system by virtue of the esoteric nature of their work, scientists were forced to confront an insistent and clamorous attack on premises that had once appeared self-evident. The Bush report was dismissed as an absurd oversimplification, and with it went much of the rationale for continued rapid growth in R&D budgets.

Especially alarming, perhaps, was that some of the deepest conflicts arose within the scientific community. Natural divisions papered over in the postwar expansion erupted: civilian researchers deplored the work of defense scientists, university environmentalists blamed industrial scientists for pollution, and so on.

The intellectual ferment in the universities raised nettlesome issues for scientists. Anti-elitism, minority access to careers, the rights of human subjects in research, and the public's right to hold private centers of power and privilege accountable were cherished notions of liberal champions of social reform. But when directed against the universities themselves, reformist ideas had worrisome consequences. They proved difficult to assimilate without upsetting long-established mores of scientific enterprise. Many university scientists, echoing the sentiments of George Ellery Hale and other prewar figures, wondered about the terms of the bargain they had persuaded the politicians and government administrators to accept. The nation would struggle for nearly a decade to devise a new approach to science policy.

Congress was already becoming more deeply engaged in science policy in the early 1960s. In fact, it rewrote the NSF's charter, moved to create new NIH institutes, and attempted unsuccessfully to establish a central Department of Science. The process of making science policy was in this respect no different from making other national policies. The postwar generation of American scholars of public administration, most of them involved in some way in wartime administration, took for granted that government would operate in a fashion akin to how it operated during the war (which was how the administrative theorists of the 1930s said it should operate). Patriotism and broad public support, deference to executive leadership, and the subordination of partial interests to the larger national interest as defined mainly by the president and his advisers were the "normal" conditions. But as the science policy debate began to break out of this established framework, the close circle of executive-branch officials and their outside scientific advisers would no longer play the dominant role.

4 ⫴ Policy Disarray, 1966–1980

THE 1960S WITNESSED a crisis in the relationship between government and science. The causes can be loosely grouped into those internal and those external to the research system. The internal factors included ambiguities and unresolved tensions in the system's original understandings that had gone largely unnoticed or had been simply ignored in the era of rapid growth. After Sputnik, science and higher education had been very well treated by both the federal government and the states. Politicians, scientific leaders, university faculty, trade association representatives, foundation executives, industrialists, and state officials saw that supporting R&D served their ends while advancing the common good. In a system with few losers, statesmanship flourished.

As the pace of expansion slowed, however, funding began to be redistributed—unequally—according to merit. Strains surfaced between faculty and administration, public and private universities, and investigators who retained and those who lost grants, not because the system failed to work but because it worked exactly as intended. It also became harder to maintain equilibrium among the various parts as the system grew in size and complexity. Small imbalances were magnified, and the disjunctions between some elements became jarring.

The most significant effects on the universities were produced by protests of the presence on campus of the Reserve Officers Training Corps and of universities' acceptance of DOD funding. These protests made many schools seem ungovernable, and outsiders were drawn into the struggle by the combatants' appeals for support. Articulate groups within a community that has always made a habit of disrespect for authority easily found ways to dramatize their grievances. The image of science as a self-governing community suffered as a result. Politicians, who may have thought that autonomy implied harmony, saw this most competitive and decentralized of human communities seemingly at war with itself and unable to agree on core values. They were happy to fill

73

the vacuum of authority, particularly when they saw that the behaviors of students, professors, and university administrators symbolized disagreements about social issues and made them good targets for political attack.

These disagreements were, of course, the most important factors affecting universities, researchers, and government agencies as well as the rest of society. The deep divisions growing out of the Vietnam War were foremost for American science as for society, but minority rights and the sharpening of racial tensions, generational disputes, the sexual revolution, the women's movement, increased consumer awareness, and the transformation of conservation into the environmental movement also affected the climate. Beginning in the late 1960s, the impact of these trends on the universities and the effects of developments within the universities on broad public debate were the dominant concerns of science policy for most of the next decade. The disputes at times became acrimonious enough to shake the foundations of the government-science partnership.

The intellectual origins of what can be referred to as the antiscience movement are diverse. One could consider the publication of Rachel Carson's *Silent Spring* in 1962 as beginning society's more critical attitude toward the uses to which science was put. Ralph Nader's 1965 book, *Unsafe at Any Speed,* was also a force. It was not antiscience as such but rather a criticism of corporate interests for misusing technology for short-sighted economic gain. Later, such thinkers as Jacques Ellul, Ivan Illich, Herbert Marcuse, and Juergen Habermas attacked the ideal of rationality and criticized the whole social structure built around science and technology. Their assaults upon modernity made them heroes of the New Left. The fears of many scientists that the Apollo program and large defense projects were absorbing too much of the nation's effort and money were also influential. The Kennedy administration's dissatisfaction with the pace of economic growth and its modest efforts to stimulate civil-sector technologies reflected the feeling that diverting technical talent from civil technology was responsible for lagging growth and a slow pace of innovation.[1] In the early 1960s Congress, too, began to call for a clear policy toward science, not quite knowing what it meant but betraying uneasiness over the terms of the postwar bargain.[2] All these forces eroded the faith in science and technology underlying the old consensus.

Perhaps 1966 best marks the transition to the more troubled period. On June 15 President Johnson delivered a speech at the launching of the

Institute of Medicine in which he called for more immediate applications from the investments in biomedical research: "A great deal of research has been done. I have been participating in the appropriations for years in this field. But I think the time has now come to zero in on the targets to get our knowledge fully applied. . . . Presidents, in my judgment, need to show more interest in what the specific results of medical research are during their lifetimes, and during their administrations." [3] By 1966, too, the National Science Foundation, the National Institutes of Health, and the Department of Defense all had launched programs to support "developing" universities that could not easily qualify for project grants based on peer review.[4] The Defense Department's Project Hindsight attempted to show that the relationship between basic research and development was more complex than delineated in the model positing that commercialization followed directly from research.[5] In the climate of the day some casual observations about basic research made near the end of the study were seized upon as proof that basic science was irrelevant to development and that agencies should focus on applied research. The same year, scientists in a Project Jason summer study session conducted under the auspices of the Institute for Defense Analyses proposed an electronic fence for use in Vietnam. When the idea proved unfeasible and was abandoned, critics denounced it as an example of misguided faith in technology.[6]

Certainly several trends had become evident by 1966. The expansion of federal R&D obligations, especially those for basic research, had slowed (they had decreased if measured in constant dollars). Insistence by politicians and the public on early payoffs from investments in research had intensified. There was a growing rift between defense scientists and those on the nation's campuses who opposed the Vietnam War and links with defense agencies. And the public had begun to doubt that unchecked scientific and technological growth would be benign.

There was, however, neither a coherent focus of opposition to science and technology nor any unifying critical premises. The forces undermining the old consensus reflected diverse and sometimes conflicting tendencies from both the political Right and Left. On the Right were those who wanted more practical knowledge, and who feared that the universities harbored leftists and allowed the teaching of subversive doctrines (many moderates agreed that a shakeout of the whole overextended research system would promote efficiency). An absurd measure to have Congress screen some 14,000 research awards annually actually passed the House of Representatives before being killed in the Senate. The measure had its counterpart from the Left in Senator William Proxmire's golden fleece awards for research boondoggles.[7]

The Left drew on traditional American attitudes of a different sort: dislike of elitism, distrust of links between corporations and universities, and admiration of spontaneity, small organizations, and nature as opposed to the dehumanizing and exploitive forces of big technology.[8] These attitudes were not confined to student radicals and alienated intellectuals; they reached back to sentiments that drew Harley Kilgore and Maury Maverick into wartime alliance and still farther back to the small farmers, shopkeepers, and artisans of a simpler past. Both liberals and conservatives could agree on certain propositions: professors should teach more, research was costing too much and often produced too little, and science had become an overblown enterprise serving its own interests and hiding abuses beneath appeals for autonomy.

More serious than the specific attacks on science, however, was the broad movement of public opinion. Even those who continued to have faith in scientific progress and defended scientists against the critics found it more difficult to believe that the nation's problems could be solved through science. The Vietnam War symbolized the dilemma: the most powerful and technologically advanced nation on earth was unable to subdue a technologically backward enemy. The events of the war seemed to say that some problems had no technological fix. The deepest aspirations of colonial peoples in a distant region eluded easy comprehension and defied the solutions proposed by systems engineering. At home the problems of racial inequality, urban decay, poverty, and discord among generations also resisted technological solution. The problems were too diffuse, entangled with jurisdictional rivalries, and fraught with controversy to permit any agency or level of government a clear mandate for action.[9]

Despite initial flurries of enthusiasm about using scientific discoveries to resolve urban problems, confidence waned quickly. The Rand Corporation set up a New York City–Rand institute to work with Mayor John V. Lindsay's urban affairs officials and discovered slippery water, which reduced resistance and enabled fire hoses to spray at higher velocities and volumes.[10] Computers helped schedule and dispatch fire trucks, keep records, and allow rapid retrieval of information. But these modest accomplishments did not address the root causes of urban problems.

For a time people believed that spillover from the space technologies would create new products and job opportunities, but the hopes mostly proved groundless because the size, scale, and cost of systems engineering ventures in space or defense programs simply could not be carried over.[11] The reality of federalism meant that there was no unified authority capable of mobilizing the resources, specifying the design of systems, or implementing proposals with complete control.

Alice Rivlin, in her influential *Systematic Thinking for Social Action*, argued the case for a rational, problem-solving approach to social ills.[12] Analytic methods and social experimentation, she urged, could be used to frame problems, test programs, and evaluate results. She had served as assistant secretary of planning and evaluation in the Department of Health, Education, and Welfare, a position created in a number of federal agencies during the 1960s to apply systems analysis, planning-programming-budgeting (PPB), and operations research—which were thought highly successful in decisionmaking for defense and space matters—to domestic problems. (The model was the office set up by Charles J. Hitch of the Rand Corporation when he moved to the Pentagon as part of Secretary Robert S. McNamara's management team in 1961.)[13] Unfortunately, the high expectations were never realized. Part of the reason lay in differences between the civilian and defense missions and organizational structures and part in misunderstandings and exaggerations of what had actually been achieved in the defense agencies. A more serious factor was the loss of faith that scientific methods offered any special or unique promise of resolving the muddy problems of social policy.

Analytic methods have been most useful when they have helped policymakers chart the best course toward an agreed objective. When the ends of policy are unclear, disputed, or too many, the analyst cannot assess the costs and consequences of alternative means. The 1960s and 1970s, especially, were a time of conflict, divisions, shifting attitudes and loyalties, and wrenching social change. The policy conflicts and underlying differences in values shook all institutions and all settled habits of thought. Society's support for science had been based on the assumption that progress in the various scientific disciplines would ultimately lay the foundation for a better life for all Americans. Social improvements of all kinds would follow when the nation's collective intelligence was brought to bear on the most pressing problems. But as Americans lost confidence in this premise, as their optimism about the future became tinged with pessimism, the foundations of society's support for science—and scientists' faith in themselves—eroded.

Challenges to the Articles of Faith

The new mood challenged the beliefs underlying the postwar scientific enterprise: the importance of basic research to the vitality of the whole system, the need to support it through a variety of sources, especially through the mission agencies, and the principle that scientific merit, as defined largely by the scientists themselves, should govern funding decisions. Within the executive branch there were demands for speeding the pace of applications from basic research. As a result, strong congressional supporters of R&D became more equivocal in their backing of basic research and critics became more vocal. Those who took little notice of science when federal R&D budgets were small were now drawn in as interested observers and persistent questioners of agency plans. The growth of federal support slowed. Universities experienced a decade of friction with the federal government. Advocates of research for its own sake were put on the defensive.

Congressional interest in R&D policy, beginning with the 1963 select committee chaired by Representative Carl Elliott and quickly followed by a second committee chaired by Representative Emilio Q. Daddario, focused on whether it was possible to achieve greater uniformity in the geographic distribution of federal contracts and grants without affecting the quality or cost of R&D. Agency responses to congressional inquiries became ritualized: "We cannot hope to obtain maximum output of research and development results unless we support and use the facilities and the scientists and engineers in our great centers of scientific and technological activity." However, "this does not mean that we are helpless in the effort to assure more widespread distribution of support for research and development. . . . [But to] build up a broader geographic base of capability . . . will require additional expenditures." Congress would find the answer unsatisfactory and would respond that the problem "was severe enough, the needs and potential benefits involved are great enough, to warrant continued attention by Congress and the Executive Branch." [14]

In September 1965 President Lyndon Johnson sought to resolve the impasse by directing agency heads to fund research "not only with a view to producing specific results, but also with a view to strengthening institutions and increasing the number of institutions capable of performing research of high quality." This statement reinforced his broader educational goals of "helping small and less well developed colleges improve their programs."[15] The directive led the National Science Foundation to expand its University Science Development program and to initiate its Departmental Science Development and its College Science Improvement programs. The Department of Defense launched Project Themes, aimed at universities not heavily engaged in federally sponsored research, and the National Institutes of Health supplemented their grants with a developmental program, the Health Sciences Advancement Award.

Yet Congress remained dissatisfied and intensified its oversight of science policy. In March 1966 Daddario introduced the first version of what would become the Daddario Bill to revise the charter of the National Science Foundation. In the Senate a companion bill was introduced by Edward M. Kennedy in October 1967. The Daddario-Kennedy Amendment, eventually signed into law by President Johnson on July 8, 1968, "changed the National Science Foundation in both form and substance."[16]

The amendment authorized the NSF to fund applied as well as basic research, designated the social sciences as eligible for support, and required the National Science Board to report annually to Congress on the state of U.S. science. The NSF's first effort to develop an applied science program was called Inter-disciplinary Research Relevant to Problems of our Society (IRRPOS) and drew on its own engineers and social scientists to define relevant research projects not fitting neatly within established disciplines. The program was criticized by Congress as inadequate, and in 1971 the small IRRPOS concept was transformed into a major effort known as Research Applied to National Needs (RANN). The NSF, in effect, traded a budget increase for program redirection.

Congressional scrutiny intensified as the NSF began to grow (a growth enthusiastically supported by NSF Director William McElroy, who announced in 1969 that he wished to increase the foundation's appropriation from $400 million to $1 billion within three years). An annual authorization cycle replaced a standing NSF authorization, which seemed to increase congressional involvement in the foundation's budget.

Congress also increasingly asserted itself in higher education policy generally, so that by 1972 the power to shape higher education policies had largely moved from the executive to the legislative branch. Unfortunately, policy for higher education and policy for supporting academic research proceeded on separate tracks.[17] Instead of producing a coherent policy, congressional power added to the disarray that had prompted greater oversight in the first place. The split between the Democratic

Table 4-1. **Funding for R&D, by Sector, 1953, 1960, 1965–87**
Millions of dollars

	Current dollars					Constant (1982) dollars				
Year	Total	Federal Government	Industry	Universities and colleges	Other nonprofit institutions	Total	Federal Government	Industry	Universities and colleges	Other nonprofit institutions
1953	5,124	2,753	2,245	72	54	19,744	10,590	8,671	276	208
1960	13,523	8,738	4,516	149	120	43,648	28,191	14,591	479	387
1965	20,044	13,012	6,548	267	217	59,351	38,532	19,384	791	643
1966	21,846	13,968	7,328	304	246	62,589	40,047	20,962	875	706
1967	23,146	14,395	8,142	345	264	64,406	40,057	22,654	960	735
1968	24,605	14,928	9,005	390	282	65,458	39,788	23,869	1,049	752
1969	25,631	14,895	10,010	420	306	64,672	37,660	25,166	1,071	775
1970	26,134	14,892	10,444	461	337	62,405	35,636	24,851	1,111	807
1971	26,676	14,964	10,822	529	361	60,385	33,966	24,387	1,212	820
1972	28,477	15,808	11,710	574	385	61,414	34,146	25,190	1,246	832
1973	30,718	16,399	13,293	613	413	62,427	33,478	26,837	1,268	844
1974	32,864	16,850	14,878	677	459	61,467	31,726	27,578	1,298	865
1975	35,213	18,109	15,820	749	535	59,883	30,986	26,679	1,302	916
1976	39,018	19,914	17,694	810	600	62,134	31,813	28,058	1,305	959
1977	42,783	21,594	19,629	888	672	63,653	32,152	29,176	1,325	1,001
1978	48,129	23,876	22,450	1,037	766	66,769	33,172	31,087	1,446	1,064
1979	54,933	26,815	26,082	1,198	838	70,077	34,271	33,198	1,538	1,071
1980	62,593	29,453	30,913	1,318	909	73,235	34,548	36,066	1,555	1,066
1981	71,840	33,405	35,944	1,520	971	76,610	35,685	38,257	1,631	1,037
1982	79,316	36,505	40,096	1,690	1,025	79,316	36,505	40,096	1,690	1,025
1983	87,204	40,671	43,515	1,881	1,137	83,891	39,097	41,896	1,805	1,093
1984	97,638	45,340	49,066	2,024	1,208	90,541	42,007	45,544	1,871	1,119
1985	107,436	51,276	52,597	2,259	1,304	96,532	46,030	47,310	2,023	1,170
1986	114,697	55,273	55,549	2,500	1,375	100,398	48,318	48,702	2,177	1,201
1987	123,050	60,350	58,570	2,700	1,430	104,345	51,154	49,693	2,286	1,212

Source: National Science Foundation, *National Patterns of Science and Technology Resources: 1987*, NSF 88-305 (Washington, 1987), table B-5.

Congress and the Republican executive branch resulting from the 1968 elections helped guarantee aggressive congressional oversight of administrative action.

The effects of the changed climate, including the student disturbances that swept the nation's campuses, on research expenditures were noticeable. From 1953 through 1967, national R&D expenditures in current dollars grew by more than 350 percent (table 4-1). The federal government increased its expenditures even more rapidly—almost 425 percent. But starting in 1967 and lasting for a decade, the growth rate was significantly slowed. Federal R&D spending declined even in current dollars for 1969 and 1970, then recovered to the levels of 1967–68 before modest increases began in 1975–76. In constant dollars, the funding slowdown was more dramatic. The federal share declined more sharply than the other components of the national effort, and outlays for basic research fell faster than total federal R&D outlays. Federal constant dollar expenditures on basic research fell 15 percent between 1968 and 1976.[18]

Federal outlays began to recover in President Gerald Ford's fiscal 1977 budget request, and basic research was singled out as needing special attention, but not until the middle of the Carter administration did it begin to receive significant increases. Greater increases followed as the Reagan administration accelerated defense R&D spending and expanded NSF and NIH budgets even during periods of severe fiscal stringency.

Another significant trend during the crisis decade was increased federal emphasis on targeted research. The trend is difficult to document, but the use of more detailed definitions of research objectives and procedures, and of contracts rather than grants, was clearly discernible. Contracts, for example, accounted for 12 percent of the NIH budget in 1967 and 26 percent in 1975, prompting an expression of concern from the President's Panel on Biomedical Research.[19] Newer agencies such as the Department of Transportation, which provided only small amounts of research funding, were mostly interested in applied research that was very narrowly focused on specific agency missions. But even such established agencies as DOD, NASA, HEW, and NSF emphasized applied research to a greater extent or exhibited less interest in mission-oriented basic research.[20]

One action important in favoring more targeted research was the Mansfield Amendment (section 203) to the Military Procurement Authorization Act of 1970, which specified that "none of the funds authorized to be appropriated by the act may be used to carry out any research project or study unless such project or study has a direct and apparent relationship to a specific military function or operation." Although the amendment was only in force for one year before the language was modified to permit a looser relationship to military needs, it illustrated the mission agencies' tendencies to define more strictly the kinds of research they considered relevant. The National Science Board felt sufficiently concerned that it issued an appeal in 1974 for agencies to increase their support for basic research potentially related to their missions.[21] The Mansfield Amendment was not motivated by lack of interest in the welfare of the universities. Rather, the liberal Democrats and moderate Republicans who supported the measure thought that they were helping universities by removing a cause for campus disturbances. Moreover, they believed it was logical to support basic research through nondefense agencies. The amendment, Rodney Nichols wrote, was significant "not because its direct effects have been great but because it is the formal expression of deeper congressional concerns, the tip of an apparently large iceberg—and it does have the stamp of the majority leader of the Senate."[22]

In the meantime, science and engineering Ph.D.s, products of expanding federal support in the 1960s, began to flood the labor market, creating a combination of reduced funding and a scientific establishment that was still growing.[23] In constant 1967 dollars, annual research expenditures per full-time-equivalent scientist and engineer employed in doctorate-granting universities decreased from $11,500 in 1966 to $8,400 in 1974.[24] Moreover, indirect costs of both private and public universities increased as a percentage of total R&D funds, further squeezing the resources available to support direct costs of research.

Few aspects of government-science relations have been so widely misunderstood or so acrimonious as controversies over indirect costs. There is no need to repeat the arguments, but the indirect costs were real. They resulted from increased expenses of departmental and university administration, maintenance of aging facilities, energy price increases, and the costs of meeting new government regulations. The consequence was to exacerbate tensions between university administrators and faculty researchers, and to complicate the relationship between universities and funding agencies.

At the start of the 1970s the federal government cut back sharply on funding for mechanisms that had previously supplemented project funding of university research. For example, in 1965 the NSF's institutional support programs accounted for 20 percent of its operating budget; by 1975 it had phased out all of them, and as of 1989 still has no institu-

Figure 4-1. **Federal Obligations to Universities and Colleges for R&D Plant, Fiscal Years 1963–82**

Millions of dollars

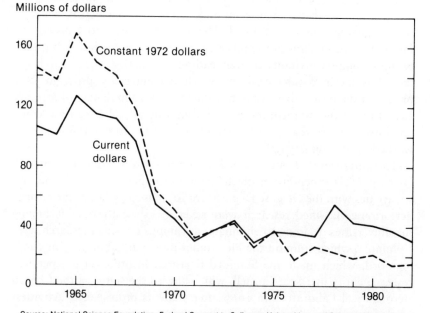

Source: National Science Foundation, *Federal Support to Colleges, Universities, and Selected Nonprofit Institutions, FY 1982*, and *Federal Funds for Research and Development*, NSF 81–315 (Washington, 1981), p. 17.

tional grants. Federal obligations for R&D plant, facilities, and large equipment in colleges and universities were also substantially reduced. They recovered briefly in 1978 but then fell again until 1984 (figure 4-1). The reluctance of state governments to provide for maintenance and support facilities exacerbated the problem. In addition, federal fellowships for graduate students were cut back.[25]

Of course, such trends affected universities and scientists unequally. Because retrenchment hit developing institutions harder than established ones, the research of many of the nation's best scientists was supported uninterruptedly. The system, in fact, worked as intended in protecting the research judged most meritorious. The most highly ranked chemistry departments, for example, increased their share of federal funds for research between 1972 and 1975.[26]

Some critics, and even sympathetic observers within the scientific community, felt that the post-Sputnik expansion of higher education had proceeded too rapidly, leading to support of some poor quality work. A pause was not altogether bad, in their judgment, so that the less productive could be screened out and overall efficiency improved. There was something to these arguments, but these years were still difficult for the universities, which felt themselves under siege. The partnership with the federal government was strained. The neglect of laboratory instrumentation and the erosion of the physical infrastructure for research threatened the long-term vitality of even leading universities. The research system thus faced a quiet crisis in which the university programs that produced the basic advances in most fields seemed in danger of atrophy. Even though the momentum generated in a decade of expansion could continue to carry the system forward for a time, thoughtful observers warned of waning strength.

The pluralism that once seemed so important was also compromised. Defense R&D expenditures on the nation's campuses declined, and university ties with the Defense Department for managing large laboratories performing classified research were severed. Especially notable among the universities deciding to eliminate on-campus classified research were Columbia, which spun off the Electronics Research Laboratory in 1967; Stanford, which made the Stanford Research Institute an independent research entity in 1970; Cornell, which altered the status of the Cornell Aeronautical Laboratory in 1969 after students protested its activities; Michigan, which converted its Willow Run Laboratories in 1972 to an independent, nonprofit research institute emphasizing environmental research; and MIT, which separated the Instrumentation Laboratory from the university in 1973 and renamed it the Charles Stark Draper Laboratory, losing a substantial management fee in the process.[27]

NASA funding at universities also dropped markedly after the completion of the Apollo mission. NIH and NSF emerged as the major supporters of many fields of academic science and engineering. NSF took over most of the responsibility for ground-based astronomy and low-temperature physics, became a major supporter of university engineering research, and in the early 1970s assumed responsibility for materials research from the Defense Department's Advanced Research Projects Agency. It also became a leading funder of oceanographic research, the antarctic research program, and work in atmospheric sciences (previously, a more pluralistic pattern of support in the oceanographic and atmospheric sciences involved the navy, air force, and NASA as well).[28] NIH support for research in the life sciences and chemistry also grew steadily.[29] As of 1975 NIH and NSF provided three-quarters of all federal funds allotted to university research, and HEW alone accounted for more than half the federal funds going to university research. Five agencies accounted for 90 percent of such funds.[30]

The clarification of responsibilities for supporting research in various fields had a number of beneficial consequences. Work in some was securely supported for the first time. But the diminished pluralism had costs as well. Investigators, especially those with a novel or unconventional approach, were no longer as free to seek alternative sources of support. The system became less dynamic: there was some tendency for researchers to avoid risks in their proposals, since a grant turned down could mean the end of a scientific career. A measure of conservatism thus crept into the system.

Promoting Civil Sector Technology

The framers of postwar science and technology policy felt little need to promote innovation and commercialization. Neither the Bush nor the Steelman reports paid much attention to industrial innovation. The Steelman report did call for constant innovation so that American industry could keep ahead of growing foreign competition, but this was to be accomplished by making sure the United States maintained overall scientific and technological leadership through a strong effort in basic and applied research and the education of high-quality scientists and engineers. Market forces ensured that research would result in products, economic growth, and jobs. Entrepreneurs rather than government planners should guide investment. The American lead in productivity after the war appeared to validate these beliefs. America's industrial history showed that the economy would generate innovations fast enough without much need for explicit policies. As Harvey Brooks notes:

> The implicit message of the Bush report seemed to be that technology was essentially the application of leading-edge science and that, if the country created and sustained a first-class science establishment based primarily in the universities, new technology for national security, economic growth, job creation, and social welfare would be generated almost automatically without explicit policy attention to all the other complementary aspects of innovation.[31]

Government policy would therefore focus on supporting research and training scientists, engineers, and technicians. Industry had an interest in patent policy but was unable to bring about a clear, uniform policy on rights to commercialize discoveries made while performing contract research for the government. Industry was, of course, not only prepared to perform large-scale R&D for the government but was eager to sell products to it. A modest level of regulation, enforcement of contractual obligations, and means for settling disputes would, as before, be an acceptable part of the relationship. The few exceptions to reliance on the marketplace, such as agricultural policy, simply underscored the nation's disregard for doctrinal purity. And even here there was a semblance of a market philosophy: public policies would support growing foodstuffs as commodities, but would stop that support once food entered the processing chain and became appropriable as a brand name product. Support for research in commercial aviation, dating from Franklin D. Roosevelt's order as assistant secretary of the navy in World War I to pool patents, and the applied research undertaken after the war by the National Advisory Committee on Aeronautics (NACA) was also critical in the development of commercial aviation. Such government support could be justified because it improved the quality of a product it was purchasing. Much the same was true of support for atomic energy research after World War II (see chapter 3).

Aside from these special cases, the first efforts to foster the commercialization of civilian technology as a matter of broad public policy began in the early 1960s. President Kennedy's economic advisers were worried about the sluggish performance of the economy and sought ways to improve the immediate economic outlook. But they realized that productivity growth was a concern that had implications going well beyond short-term stabilization policies. In the summer of 1962 the president established the Cabinet Committee on Growth to address these issues, drawing on staff from the Council of Economic Advisers and the cabinet departments. The president's science advisers were at work along parallel but somewhat narrower lines on the Civil Technology Panel, which drew members from the President's Science Advisory Committee, the Office of Science and Technology, and the Council of Economic Advisers. The panel summed up its views in a 1962 report to the president, *Technology and Economic Prosperity*, in which it urged new steps to strengthen the nation's technological leadership and to boost production.[32]

The OST report clearly reflected the Kennedy administration's thinking on the challenges posed by economic slowdown and possible erosion of technological leadership. Market mechanisms had generally worked well in fostering innovation, but market imperfections were becoming

evident. Some firms were short-sighted and lacked the expertise to estimate the benefits of investing in new technologies. Others could make
accurate, short-run benefit-cost calculations but could not take into account industrywide or societywide effects that would come from the
adoption of new processes or products. This problem was exacerbated
because the federal government, by monopolizing many of the nation's
best scientists and engineers in the defense and space programs, diminished the pool of talent available to civil technology. Committing resources to research and development in defense and space science was
appropriate but contributed little directly to economic growth. More efforts were therefore needed to stimulate civilian technology and expand
the pool of talent so that the government's own programs would not be
a brake on economic expansion.

Although their emphases differed slightly, the two reports were the
first to argue that the needs of defense and civil technology were in conflict. This crack in the postwar consensus was partially closed by familiar
calls for expansion, particularly for awarding more advanced degrees
in the sciences and engineering. But educational efforts would not be
enough: the administration groped toward a program combining research, extension services, and demonstration—as in the support of agriculture—to stimulate lagging industries. Because textile manufacturing, coal mining, and housing construction seemed good cases for special
assistance, the administration devised the Civilian Industrial Technology
program, the first of what was to be a series of initiatives to stimulate
innovation in the civilian economy. The CITP was defeated in Congress,
however, because of intense opposition from the very industries intended
to be the program's beneficiaries. The housing industry did not believe
its research and development expenditures were inadequate, nor did it
believe it suffered from market failure or that there was a direct link
between more research and increased business efficiency.[33] The administration also failed to make its case to the coal and textile industries. And
Congress was not sufficiently concerned about long-term growth prospects to force a departure from the prevailing assumptions.

Wanting to speed the progress of research from laboratory to application, the Johnson administration took up the challenge of civilian technology anew. In 1965 the State Technical Services Act, an effort to modify
the Kennedy CITP to take into account the criticisms raised by industry
and other opponents, was signed. Instead of providing for federal government research on industry's behalf in what the government decided
were lagging sectors, the act stressed the exchange of technical informa-

tion and consultation among industry, the universities, and state governments. Donald Hornig, President Johnson's science adviser, considered that the major function of the new program was to provide information on the best practices in manufacture through state and local representatives familiar with both local conditions and manufacturing technology.[34] Advanced technologies were assumed to be commercially feasible, an assumption industry in general did not share. Industry saw public assistance for R&D as contributing relatively little toward reducing the total costs of innovation; most costs were downstream from research and the early phases of development. The naive public belief that R&D is all there is to innovation, in industry's view, simply increased the political pressure to make huge investments in new technologies that might not pay off. This reaction was also to be characteristic of industry's later skepticism of the mandates to use "best available technology" that appeared in environmental and occupational health and safety legislation.

Because STS was a modest program, concern about civilian innovation continued to mount. The Office of Science and Technology advocated more extensive government efforts to force the pace of innovation. The Urban Institute was created in 1965, with William Gorham of the Rand Corporation as its president, to apply systems analysis—techniques of analysis used in the defense and space agencies—to a host of urban ills. Systems engineering and analysis would first clarify policy choices, then lead to restructuring markets and state and local government procurement practices to permit aerospace and other research-intensive industries to devise and sell advanced technologies to solve local problems. This could mean, for example, cheaper construction materials for housing, advanced waste disposal systems, high-speed transit, and, in general, a faster pace of innovation. There was also an emerging concern about future unemployment in the high-technology industries as the manned space program and procurement of advanced defense systems slowed. The obvious solution, it seemed to many administration strategists, was to help rechannel scientists and engineers from aerospace to less research-intensive sectors.[35] There was also a theory in vogue that radical innovation would be most likely to occur through "invasion"— the entry of new companies into a traditional business area.

In 1966 Charles Zwick, director of the Bureau of the Budget, chaired a White House panel appointed to evaluate the recommendations of the various committees dealing with innovation and economic growth.[36] The panel found little evidence that the private sector neglected innovation and no compelling justification for government to provide technical assistance. Thereafter President Johnson's science and economic advisers remained split over whether more public assistance was needed to speed commercialization of R&D. The science advisers urged long-range R&D programs in transportation, housing, pollution control, and water resources development; and they wished to strengthen engineering schools and university applied science programs. They also wanted to remove barriers to innovation, a harbinger of the deregulation movement of the late 1970s and 1980s. This was somewhat paradoxical, for the nation was also moving toward more intensive regulation in health and safety matters.

In January 1967 the Commerce Department released a major report by a group of influential citizens in collaboration with the Commerce Department's Civilian Technology Advisory Board and the office of J. Herbert Holloman, assistant secretary of commerce for science and technology. *Technological Innovation: Its Environment and Management* (the Robert Charpie report) called for creation of a technical advisory body to study the impact of antitrust and regulatory policies on innovation and to advise government agencies on steps to ease regulatory constraints.[37] The Commerce Department had established the Civilian Technology Advisory Board to provide technical advice to Holloman and the secretary of commerce (the board was disbanded at the start of the Reagan administration).

A related concern with government procurement practices and their impact on the pace of innovation was also important. With OST's support, the Office of Urban Technology and Research in the Department of Housing and Urban Development sought to standardize the purchasing policies of state and municipal governments. If housing codes could be harmonized and larger markets created that would justify R&D investments, it might be possible to develop advanced materials or modular designs that would make housing construction cheaper.

In contrast to the science advisers' advocacy of technological solutions to problems, the president's economic advisers, especially the powerful institutional voice of the Budget Bureau, sought to block even small programs out of fear that they might grow into large, wasteful, bureaucratic interferences with the marketplace. The two sets of advisers agreed on other issues of science policy, particularly on efforts to control defense and space expenditures, but given their differences in perspective, no clear objectives for civilian technology or the commercialization of research were possible.

The Nixon administration carried the challenge to the postwar consensus much further and formulated more ambitious plans to foster innovation before eventually succumbing to policy incoherence and internal divisions like those affecting technology policy in the final years of the Johnson administration. There was little that was totally new in the Nixon administration's approach. It mainly took concepts already under discussion and program elements already in place and emphasized the ideas through expanded programs. President Nixon, like President Johnson, considered federal R&D a way of helping to solve social problems. The administration's task was to redirect existing government programs where necessary to reflect new priorities. Nixon sought to explicate his science policy in a 1972 presidential message.[38] This was the first such presidential message, and it came partly in response to continuing congressional pressures for a clearer statement of the administration's views. It gave prominence to problems with the environment, health, energy, and transportation and attempted to show how the various elements of the nation's total R&D effort were addressing these needs. But by the time the message appeared, the divisive tendencies within the administration were growing and the prospects for a clear, central voice in science policy were diminishing.

An early Nixon administration initiative was Operation Breakthrough, a program advocated by Secretary of Housing and Urban Development George Romney to improve research on housing construction.[39] This effort sought to provide visibility and momentum for programs begun in HUD's Office of Urban Technology and Research during the Johnson administration. The Nixon administration also expanded the State Technical Services program. Transportation Secretary John Volpe took a strong interest in mass transit problems, and his department constructed a demonstration high-speed rail line in Morgantown, West Virginia, as a prototype. The creation of the Transportation Systems Center in Cambridge, Massachusetts, was a related initiative.

In 1971–72 the administration launched an ambitious effort to identify new technological opportunities and brought in William M. MacGruder, an aerospace engineer, to coordinate the program. The New Technologies Opportunities program rested on the idea that many of the promising technologies developed under government auspices could be transferred to the private sector and find wide application once their feasibility had been demonstrated. Agencies were accordingly asked to suggest ideas for inclusion on a list of priority projects for advanced development and early commercialization. The agencies responded with many technologically feasible but costly proposals, including new nuclear power systems for commercial shipping, high-speed rapid transit, deep-water explorations, and offshore ports for large tankers.[40] These ran up against opposition from the Office of Management and Budget and strong technical criticism from the OST and the President's Science Advisory Committee. The program eventually stalled. Fiscal constraints were partly to blame, but fundamental ideological impulses surfaced once again. Critics objected to the degree of intervention in the marketplace that would be involved: it seemed inappropriate and unwise for a Republican administration to subsidize the development costs of private technologies. Such efforts might reduce the incentives for privately funded R&D, thus displacing resources from the commercial to the government sector. The administration eventually approved a small number of experimental programs in several agencies, including the National Bureau of Standards and the National Science Foundation, as well as some small research programs.

Meanwhile, President Nixon was growing disenchanted with his scientific advisers. Opposition to administration policies on the supersonic transport and the Anti-Ballistic Missile Treaty, and in particular the publicly expressed opposition by several members of the President's Science Advisory Committee, contributed to his decision to reorganize the advisory system. The pro forma resignations of PSAC members preceding his second presidential term were accepted and no new members were appointed; thereupon presidential science adviser Edward E. David, Jr., resigned to return to industry. Reorganization plan number one of January 26, 1973, issued just before the expiration of the president's reorganization powers, provided for abolishing or transferring out of the executive office various components of the science advisory system. It has been widely assumed that after this step there was no effective voice in the Nixon administration for science and technology policy and that the various federal R&D agencies pursued their own priorities (as modified by the budget constraints and policy directives set forth through the OMB). This conventional wisdom does not, however, fit the facts; at least it overlooks the considerable thought that preceded the decision to abolish the PSAC. More seriously, the argument that Nixon abolished the science advisory system slights the effective activities of H. Guyford Stever, the NSF director, and a strong NSF staff from January 1973 until the presidential science advisory system was reestablished at the end of the Ford administration.

Budget officials were particularly delighted with the reorganization, arguing that there was no need for a separate staff focusing on science and technology policy in the Executive Office of the President: policy could be coordinated effectively through the normal budget and planning functions of OMB, a point argued by George Shultz, OMB director, in an appearance before the National Science Board in the spring of 1973. The interagency committee chaired by NSF Director Stever, which was continued after the reorganization, also provided coordination.

One final aspect of civil technology policy that received presidential interest in the Nixon administration and into the Ford administration was energy. On June 29, 1973, Nixon appointed John A. Love as assistant to the president for energy and director of the new Energy Policy Office in the Executive Office of the President.[41] The president proclaimed an ambitious goal of achieving energy independence, and thereafter energy R&D expenditures increased dramatically.[42] A variety of nonnuclear technologies, as well as an expensive nuclear fusion program, were emphasized. Along with this came organizational initiatives that sought to make the resolution of energy problems a national priority.

On December 31, 1974, Congress passed the Federal Nonnuclear Energy Research and Development Act, which abolished the Atomic Energy Commission and divided its functions between the new Energy Research and Development Administration (ERDA) and the Nuclear Regulatory Commission (NRC). In addition to taking over nuclear R&D activities of the Atomic Energy Commission, ERDA was given significant new nonnuclear R&D responsibilities, particularly a federal program of assistance to and demonstration of commercially promising energy technologies. The program was never precisely defined and gave rise to many of the same disputes that had plagued demonstration efforts in housing and rapid transit. The general assumption was that by demonstrating the feasibility of a system the government would pave the way for entrepreneurs who would invest their own money and take the additional steps to market the product or service. As long as energy prices remained high, alternative fuels, energy conservation systems, and other related technologies seemed commercially feasible.

In the Ford administration the OMB, responding to the president's preference for market mechanisms to support research, sought unsuccessfully to curb the growth of demonstration programs. It tried to limit federal support to those cases in which the government was the customer and primary user of the R&D, in which market failures or special conditions prevented adequate private investment in the research, or in which a strong national consensus existed on some urgent national need that was inadequately met by the private sector.[43] These criteria justified defense and space research expenditures, basic research, and the rapidly growing energy effort, so that they did not affect the pattern of R&D expenditures. The OMB did, however, prevent the growth of large new programs and phased out some of the smaller civil technology programs begun under President Nixon. Deregulation, especially in energy, transportation, banking, and communications, also became important to President Ford's strategy. Realistic, market-driven pricing and easier entry into and exit from regulated sectors, administration strategists believed, would give a significant boost to innovation.

The slowdown in funding for basic research troubled some members of the administration, notably Vice President Nelson Rockefeller, who urged the president to consider reestablishing a presidential science advisory system (and who argued unsuccessfully for a $100 billion energy and critical materials program). On December 21, 1974, Ford authorized Rockefeller to study the possibility of reviving the White House science advisory system. This led eventually to passage of the National Science and Technology Policy, Organization, and Priorities Act of 1976. The act directed a review of the current federal science, engineering, and technology effort, recreated the Office of Science and Technology Policy under the supervision of a director (who would also be the science adviser), and established an intergovernmental panel to advise the director on ways to improve technical capabilities in state government. The OMB grudgingly supported the re-creation on the premise that additional White House staff expertise and political clout might help discipline the agencies, in particular the sprawling energy bureaucracy.

The Carter administration marked both the high point and the beginning of the movement away from the policy of extensive government involvement in commercializing R&D. The Carter years witnessed the 1978–79 oil shock, the sharpening of fears about the loss of markets abroad and import penetration and job loss at home, and calls for a strategy to reindustrialize America. Energy policy remained a priority, but the administration also groped toward a broader strategy in support of civilian technology. At the same time, movement toward regulatory reform, increased emphasis on support for basic research, and the application of more stringent market tests for large demonstration projects adumbrated the priorities that were to guide the Reagan administration's science and technology policies.

The Carter administration very early considered that restoring productivity growth and encouraging industrial innovation were important goals, and it undertook a broad review of what government could do to help. The result was a program similar to Nixon's New Technologies Opportunities but much larger. Frank Press, Carter's science adviser, suggested the review and worked closely with Domestic Policy Adviser Stuart Eisenstat, Secretary of Commerce Juanita Kreps, and Assistant Secretary of Commerce Jordan Baruch, whose office conducted the study.[44] The review involved some twenty government agencies and hundreds of outside groups and individuals. The advisers concluded that not only removing barriers to innovation but also positive executive action and legislation were necessary. A presidential message to Congress in October 1979 advocated expanding the government's efforts through the National Technical Information Service to transfer to industry the technical know-how generated in universities and government laboratories under government grants or contracts. The message proposed increasing government R&D for technologies of particular interest to industry, especially generic technologies such as welding and joining, corrosion prevention, and robotics that underlie several industrial sectors, and technologies designed to help small business comply with environmental, health, and safety regulations. It recommended expanding the NSF program to foster university-industry cooperative projects, including payments to small business to help defray the costs of their participation, and strengthening the patent system by establishing a uniform government policy with respect to patents obtained under government-sponsored R&D projects in private institutions. It also proposed clarifying antitrust policy regarding research cooperation among firms and issuing an annual guide to changes in policy by the Justice Department, expanding the NSF Small Business Innovation Research program that provided funding to small companies, and establishing several state and regional corporations for industrial development to assist high-risk innovation.[45]

In addition the administration launched several high-profile, industry-university-government cooperative programs designed to assist specific industries. The Cooperative Automobile Research program (CARP) supported research on combustion, structural mechanics, materials, and catalysis that would help the auto industry design cars for the 1990s and improve manufacturing processes. The five major U.S. manufacturers reluctantly agreed in principle to join in the effort, and Congress approved modest funding for fiscal year 1981. The program expired with the Carter administration. The Ocean Margin Drilling program, designed to provide a better understanding of the continental shelf margins through deep-sea drilling and exploration, failed when the oil companies reneged on their support as oil prices began to fall.

Representatives of industry who participated in the Carter administration's review of domestic policy considered many of the initiatives less significant than broad efforts affecting economic policy, particularly federal tax policy. They urged accelerated depreciation allowances and other tax measures as the most effective means to rebuild America's industrial base and encourage innovation. The administration responded with an economic revitalization program announced by the president on August 28, 1980, which was designed to follow up on the earlier measures: accelerated depreciation and a simplified schedule for business taxes to encourage investment in new plant and equipment, a partially refundable tax credit for depressed businesses with no earnings but with investment needs (which created the possibility of a market in unused tax credits), further assistance to small business, and various export promotion measures. Frank Press tried to include a research and development tax credit in the package but failed.[46] The tax credit and accelerated depreciation were, however, included in the Economic Recovery Tax Act of 1981, passed after the Reagan administration took office.

Administration priorities in promoting commercialization of new technologies were also reflected in energy programs, including the Ocean Thermal Energy Conversion Research, Development, and Demonstration Act of 1980, the Magnetic Fusion Engineering Act of 1980, and the National Materials and Mineral Policy Research and Development Act of 1980. Congress advanced its own initiatives, such as the Stevenson-Wydler Technology Innovation Act of 1980 to push the administration toward more aggressive action. The act established the Office of Industrial Technology in the Department of Commerce, provided financial assistance to universities for new centers to study industrial technology, mandated the transfer of technological innovations from federal laboratories to state and local governments and the private sector, and called for creation of a technical advisory board to the secretary of commerce. H.R. 6910, a bill to create a National Technology Foundation to take over NSF's applied research functions and focus on technology development, was also introduced in 1980 but failed to pass. It reflected, however, widespread dissatisfaction in Congress with NSF's activities in applied R&D and was the precursor of continuing congressional proposals for more centralized federal organization of technological activities.[47]

The Carter years were also notable for the Chrysler and Lockheed loan guarantee programs and for debate on the reindustrialization of America that was to continue into the Reagan administration. Although the rhetoric became more Republican ("competitiveness" instead of "reindustrialization"), the issues in the debate were similar.[48] The Carter administration began to revise interventionist policies when Press urged increased budget support for basic research. By the middle of Carter's term, basic research budgets were once again growing.

Increasingly, large projects demonstrating energy-efficient technologies were scrutinized by the White House science office and the OMB. The energy crisis meant that funding for them was still significant, but budget priorities had begun to change and resembled those of the Bush report (and ones that would be embraced by the Reagan administration). That the changes did not correspond in any pat fashion with the tides of electoral politics is clearly evident from the controversies surrounding control of technology and the role of science in the regulatory process. The Carter administration occupied a transitional role, inheriting previous administrations' strong emphases on protecting the environment but also paving the way for regulatory reforms.

Social Control of Technology

On November 16, 1960, President Eisenhower's Commission on National Goals issued *Goals for Americans*. Reflecting the views of eleven distinguished Americans, the report called attention to fifteen goals that would require national action in the ensuing decade. Improving environmental quality, health, and safety and related matters were not among them. Within a few years the climate of opinion had changed dramatically. Beginning with legislation in the mid-1960s to protect water and air quality, environmental concerns became national priorities.[49] Within another few years, escalating concern led to stiff protection measures and to a new framework for regulation (table 4-2). The emotional peak of the environmental movement may have been reached on April 22, 1970, when millions of Americans celebrated Earth Day. Intense legislative and lobbying activity followed.

The early 1970s also witnessed significant steps to control military technology with the Anti-Ballistic Missile Treaty and the Strategic Arms Limitation Treaty (Salt I). When Congress canceled the supersonic transport program in May 1972, environmental concerns for the first time helped defeat a major advanced-technology system. In October 1972 Congress added a new means to review rapid technological advances by creating the Office of Technology Assessment.[50]

The establishment of the OTA represented an attempt to control technological change, an attempt characteristic of this phase of postwar sci-

Table 4-2.   **Environmental Protection Statutes Enacted in the United States, 1969–76**

Statute	Year	Statute	Year
National Environmental Policy Act (P.L. 91-190)	1969	Noise Control Act (P.L. 92-574)	1972
Clean Air Amendments (P.L. 91-604)	1970	Safe Drinking Water Act (P.L. 93-523)	1974
Resource Recovery Act (P.L. 91-512)	1970	Toxic Substances Control Act (P.L. 94-469)	1976
Occupational Safety and Health Act (P.L. 91-596)	1970	Federal Insecticide, Fungicide, and Rodenticide Act (P.L. 92-516)	1976
Consumer Product Safety Commission Act (P.L. 92-573)	1973	Resource Conversion Recovery Act (P.L. 94-580)	1976

ence policy. Yet in contrast to environmental legislation, the OTA act sought to balance the positive and negative effects of introducing new technology. The National Environmental Policy Act, for example, which had required environmental impact statements, made introducing new technologies difficult, but the OTA act recognized that old technology was often the cause of harmful effects that could be fixed by more efficient and cleaner technologies. The legislative history, however, did not outline what the functions of the office would be. Some of the early rhetoric suggested that technological advance would no longer occur unhindered and that some innovations might be subject to approval before wide use would be allowed. The office would attempt to assess potential consequences of innovations as a guide to legislative action. Only the Food and Drug Administration, which tested drugs for safety and efficacy before allowing their use, could assess innovations in the manner intended by the OTA legislation, but while the FDA's jurisdiction was sharply limited, the OTA could presumably scrutinize all technology.

This grandiose concept was rejected by a pragmatic group of the OTA's initial staff, who, along with congressional and scientific allies, sought to define a more modest scope of activity. OTA's first director, former representative Emilio Q. Daddario, belonged to the pragmatists. When the OTA came under early attack as a campaign arm of Edward Kennedy in a potential 1976 presidential bid and as a launching pad for politically motivated criticisms of President Nixon, Daddario devoted his strongest efforts to helping the office survive. In subsequent years, especially under its third director, John Gibbons, the OTA became a respected source of advice to Congress. It enjoyed some leeway in responding to congressional requests, usually by seeking a consensus among congressional committees on the desirability of undertaking an assessment, but its name still evokes the spirit of strict regulation.

The change in attitude toward technology represented by the establishment of the OTA most clearly distinguishes the second from the first phase of postwar science policy. No one could any longer write of technology in the rhapsodic vein of the science editor of the Scripps Howard newspaper chain in the 1950s, when he predicted that chunks of uranium would be mounted on towers so that "No baseball games will be called off on account of rain. . . . No airplane will bypass an airport because of fog. No city will experience a winter traffic jam because of snow."[51] It is not the purpose of this study to describe in detail this change in attitude or explore specific controversies over disposing of radioactive wastes, ensuring clean water, banning artificial sweeteners, or preventing acid rain. Instead I want to describe the general shape of the change. The movement toward self-consciously attempting to control technological innovation and its effects occurred in four stages: activation of elite opinion, mobilization of interest groups, sustained attention from news media, and reinforcement and legitimation as a public policy concern through actions of government and political champions. These stages are not wholly distinct; initial actions help to trigger new cycles of awareness, interest group activity, media attention, and government action. Nonetheless this scheme outlines how the policy agenda was redefined between the mid-1960s and the end of the 1970s.

Activating Elite Opinion

In *The Chemical Feast* (1970) Ralph Nader alerted people to the dangers of uncontrolled technological innovation, depicting the chemical revolution as threatening to engulf society in a flood of uncontrolled substances. A year later in *The Closing Circle*, Barry Commoner cast industry not merely as an unwitting polluter but as a villain actively creating environmental hazards. The earlier uncritical celebration of the rapid and free progress of technology had now become a pessimistic belief in an unprecedented peril to the health of future generations.

Mobilizing Interest Groups

As well as being a prominent expositor of the dangers of technological innovation, Ralph Nader became an activist important in mobilizing interest groups. His rise, and that of the environmental movement, was assisted by astonishing corporate blunders. When Nader testified before a Senate subcommittee investigating auto safety, General Motors executives hired private detectives to follow him and try to entrap him in sexual liaisons with prostitutes.[52] When the facts became known, the GM chairman was forced to apologize publicly, and the company paid a settlement to Nader to resolve a suit for invasion of privacy. These funds helped the crusader launch a public interest organization that galvanized the environmental movement.

The Storm King controversy in the Hudson River Valley outside New York City was another key episode in this phase. Starting as a local protest by neighborhood and traditional conservation groups against the plans of Governor Nelson Rockefeller and his brother Laurence for a pumped storage project adjacent to the family's Pocantico Hills estate, the initial protest quickly gave rise to a sophisticated coalition fighting the use of nuclear power (despite the fact that Storm King was not a nuclear power station) and eventually blocked the project.[53] The national press and television evening news gave the controversy extensive coverage. Traditional conservation societies, which had been concerned mainly with protecting the rural countryside, now became more aggressive lobbyists interested in broader environmental concerns. New umbrella-type national organizations such as the Environmental Defense Fund and the Natural Resources Defense Council were also created.[54] The movement became a profusion of national and regional, single-issue and multi-issue, professionalized and amateur groups that temporarily united in the 1970 Earth Day celebration. Afterward they continued as a loosely integrated coalition of interest groups.

Sustaining Media Attention

To become effective, lobbyists require sustained public, and therefore media, attention. From the formation of Nader's Raiders and the Storm King controversy in the 1960s through the 1970s, the news media covered environmental issues extensively. Dramatic events, such as the kepone crisis in the Chesapeake Bay and pollution by PBBs in Michigan that required the slaughter of farm animals, captured the attention of millions of Americans. The activities of interest groups became part of the news, along with the reactions of officials to the various proposals to protect the environment, health, and safety. Environmental activists were clearly more successful than their opponents in capturing attention and framing issues because the climate of opinion had been prepared by environmental theorists, because they outorganized and outmaneuvered opponents, and because they had a good case.

The Rise of Political Champions

The final element in the success of the environmental movement was the support of public officials who encouraged the activists.[55] The dynamics are illustrated in the evolution of policies to cope with water and air pollution. On June 30, 1948, Congress passed the Water Pollution Control Act, giving the federal government the authority to conduct investigations, research, and surveys related to improving the nation's waters. The act, the product of a few dedicated officials, was misnamed in that the federal government was given only very limited powers to control water pollution, the primary responsibility for which was left with the states.[56] Eight years later Congress enacted the Water Pollution Control Act Amendments of 1956, which provided federal grants for constructing municipal water treatment plants and established procedures for federal enforcement actions against companies discharging pollutants into rivers and navigable waterways. Amendments that modestly strengthened these powers were added in 1961. As environmental issues became more salient, the climate was created for more extensive action; spurred by champions of environmental protection in the executive branch, Congress passed the Water Quality Act of 1965 to strengthen enforcement powers still further and provide for ambient standards for interstate waters. Momentum was maintained through the adoption of strengthening amendments in 1966 and 1970. A major breakthough came with the far-reaching regulatory steps in the 1972 Water Pollution Act Amendments.

Concerns over air pollution came slightly later, but rapidly rose to public prominence, resulting in important legislation sooner than the 1972 Water Act. The process, however, was similar. A few people operating almost invisibly within the government slowly built support for a modest program of research and sought congressional backing. Their efforts paid off when Congress passed the Air Pollution Control Act of 1955, authorizing for the first time a federal program of research, training, and demonstration projects relating to air pollution control. In 1963 Congress adopted the Clean Air Act, giving the federal government some enforcement powers through the convening of conferences similar to those allowed in the Water Pollution Control Act amendments in 1956. In 1965 the Motor Vehicle Air Pollution Control Act gave the Department of Health, Education, and Welfare the power to prescribe emission standards "as soon as practicable." The Air Quality Act of 1967 authorized HEW to oversee the establishment of state standards for ambient air quality and set national standards on automobile emissions. These actions set the stage for the Clean Air Amendments of 1970, which greatly expanded the federal powers to set and enforce standards for ambient air quality and established stringent new emissions standards for automobiles. Significant battles occurred in implementing the legislation and settling legal disputes.[57]

Thus in both cases a few voices sought to dramatize a cause already represented by a minor government program. It is rare that public policies spring full-blown without precursors. Policy entrepreneurs pick up ideas by theorists—sometimes they have inspired the theorists in the first place. Activists outside government begin to attract media attention, enabling insiders to make proposals for wider action. With effort comes modest success. While thus energizing others with an official stake in the policies, the entrepreneurs become in the eyes of the public a force to be reckoned with. Consequently they take on an even more ambitious agenda. Efforts to control technology persisted for nearly a decade (and left a strong imprint and many still unsolved policy disputes). Finally, the interplay of ideas, interests, institutions, and individuals proposing new policies blunted their momentum, modified some premises, and partially reversed the direction of environmental regulatory policies.

The Loss of Technological Advantage

The crisis in government-science relations came at a time of especially dramatic changes in international affairs that inevitably created further complexity for a policy process already rife with political and intellectual controversy. The Steelman report had warned of heightened economic competition from the industrialized nations once postwar reconstruction was over. Few foresaw in 1947, however, that the American market would not always provide a safe haven. By the 1970s companies from Japan, Europe, and the newly industrialized nations had made stunning inroads into American markets in what was termed a new industrial competition.[58] American firms no longer could compete on familiar, comfortable terms merely with each other, and they lost significant shares of the market in autos, cameras, stereos, television sets, steel, machine tools, and microelectronics to relentless foreign producers. Increas-

Table 4-3.   **R&D Expenditures for the United States and Selected Countries, 1969, 1979**
Billions of current dollars unless otherwise specified

Country	Total R&D		Private industry R&D	
	1969	1979	1979	Percent of total
United States	25.6	55.0	25.3	46.0
West Germany, France, United Kingdom	8.3	39.0	19.3	49.5
Japan	3.0	19.3	11.4	59.0

Source: Harvey Brooks, "National Science Policy and Technology Initiative," in Ralph Landau and Nathan Rosenberg, eds., *The Positive Sum Strategy: Harnessing Technology for Economic Growth* (Washington: National Academy Press, 1986), p. 136.

ingly integrated capital markets helped to complete the movement toward a global economic order.

Simultaneously America lost its seemingly effortless technological superiority. The cause, some commentators have lamented, was that U.S. expenditures on civilian R&D fell behind those of Europe and Japan, but this is incorrect. In absolute terms, U.S. expenditures continued to dwarf those of all other Western nations (table 4-3). Even if defense R&D is subtracted from total U.S. expenditures on the theory that there is no spin-off from military research and it is assumed that most privately financed industrial R&D is for commercial purposes, the United States still has a dominant position.

The absolute level of U.S. civilian R&D continues to be the highest in the world, even though the investment of other nations has grown relative to that of the United States. These numbers do not, of course, imply anything definite about the economic significance of the expenditures. Diffusion of the results of R&D does not occur automatically, so one cannot infer readily how much a given investment will contribute to output. Diffusion may also occur outside a country's borders, especially with basic and generic applied research, so that the appropriability of research results may not be easily predictable.

The ratio of civilian R&D to GNP shows that Germany and Japan have increasingly outperformed the United States and that France has steadily closed the gap (table 4-4). The ratio might suggest that other nations are deriving more economic benefit from their R&D expenditures, but there is little reason to think that the ratio is a better measure than the aggregate level of R&D expenditures. Certainly this ratio alone cannot be construed as proof of a presumed loss of U.S. innovative capacity.

Table 4-4.   **Estimated Nondefense R&D Expenditures as a Percent of GNP, by Country, 1975–85**

Year	France	West Germany	Japan	United Kingdom	United States
1975	1.46	2.08	1.95	1.55	1.63
1976	1.44	2.01	1.94	n.a.	1.62
1977	1.44	2.01	1.92	n.a.	1.61
1978	1.41	2.10	1.98	1.61	1.63
1979	1.42	2.27	2.08	n.a.	1.69
1980	1.43	2.30	2.21	n.a.	1.79
1981	1.50	2.34	2.37	1.72	1.81
1982	1.63	2.48	2.46	n.a.	1.88
1983	1.69	2.43	2.60	1.60	1.87
1984	1.76	2.41	2.59	n.a.	1.86
1985	1.85	2.53	2.75	1.71	1.86

Source: National Science Foundation, *Science and Engineering Indicators, 1987* (Washington, 1987), p. 237.
n.a. Not available.

When one concentrates on output, however, there can be no complacency. A variety of measures show that U.S. performance has slipped relative to that of its competitors. Consider the balance of trade in high-technology products, a frequently used measure of competitiveness and innovative capacity. The classification of exports into high-technology and low-technology is usually based on the ratio of R&D expenditures to sales or value added or on the number of scientists and engineers per 1,000 workers. The NSF defines technology-intensive industries, for example, as those having 25 or more scientists and engineers per 1,000 workers and spending at least 5 percent of sales on R&D. For high-technology industries overall, the largest U.S. firms held 79 percent of world markets (including the U.S. domestic market) in 1959 but only 47 percent in 1978. The market share of Japanese firms had, meanwhile, increased fourfold.[59] Table 4-5 shows the decline in market shares for the largest U.S. companies in four high-technology industries.

The U.S. share of world exports of R&D-intensive goods declined from 31 percent in 1962 to 21 percent in 1977, while the Japanese share rose from 5 percent to 14 percent.[60] However, the decline in export share is less than the decline in the U.S. share of world GNP in this period, so that the significance of the statistic is unclear. It may be that technology-intensive exports have continued to be a strength for the United States, at least in comparison with low-technology products, for which the decline in U.S. competitiveness has been most evident.[61]

Other measures also underscore the challenge to U.S. technological

Table 4-5. **Worldwide Market Shares of Largest U.S. Companies, Selected High-Technology Industries, 1959, 1978**
Percent

Industry	1959	1978
Drugs and medicine	61.6	35.0
Chemicals	66.3	31.9
Electronic products	75.6	46.9
Aircraft and parts	78.3	53.2

Source: International Trade Administration, *An Assessment of U.S. Competitiveness in High Technology Industries* (U.S. Department of Commerce, 1983), table 27.

leadership. American productivity growth from 1960 to 1980, as measured by the output per employee in manufacturing, was less than in many other industrialized countries. GNP per employed civilian worker, adjusted for relative purchasing power, remained the highest in the world (Japan's GNP per employed worker was only 75 percent that of the United States).[62] But more relevant, perhaps, is productivity in key industries, and here the Japanese, according to a 1981 assessment, overtook the United States not merely in the rate of increase but in absolute levels. Japanese productivity levels for 1979, for example, were 108 percent higher than those of the United States in steel, 11 percent higher in general machinery, 19 percent in electrical machinery, 24 percent in motor vehicles, and 34 percent in precision machinery and equipment.[63] There is clearly a correlation between these gains and Japanese success in penetrating American domestic markets.

The United States has continued to lead all major industrial countries except the Soviet Union in the numbers of scientists and engineers engaged in R&D per 10,000 civilian workers, but the margin over Japan and West Germany narrowed greatly in the 1970s, and Japan was producing more engineers from undergraduate programs. Considering that 50 percent of all U.S. scientists and engineers worked on military projects, the slim American lead in the ratio was hardly reassuring.[64] Perhaps more significant, the so-called technical literacy of the U.S. labor force apparently lagged behind its competitors, particularly Japan and West Germany. The mathematics and science proficiency of U.S. high school students was inferior to that of students in these major competitors. Another frequently cited statistic indicating the erosion of U.S. technological capacities is the 38 percent drop between 1970 and 1982 in the number of patents granted to Americans by the Patent Office, while the number granted to foreign inventors doubled to 26 percent of all patents.[65] Meanwhile, the number and share of foreign patents granted to

U.S. inventors also declined.[66] The economic significance of patents is, however, difficult to assess.

Nonetheless, taken together these measures point to dramatic economic changes that have shaken the postwar faith in American technological supremacy. What remains unmistakable is that the United States no longer monopolizes advanced technology. Some longstanding American weaknesses, hidden or forgotten in the great postwar expansion, have been exposed as other nations have developed, and the technical advances of other nations have manifestly intensified the competitive challenge America has faced since the 1970s.

Science in the State Department

All these factors were reflected in the policy debates that marked government-science relations in the 1970s. The nation had to confront not only domestic changes but also difficult worldwide trade and foreign policy developments. Protectionist sentiments, hardly in evidence when American companies enjoyed seemingly unassailable positions, revived as communities suffered dislocations and job losses. U.S. national security seemed threatened by the leakage of technology from allies to Eastern Bloc nations: when the United States dominated most of the sensitive technology, there was much less occasion for friction with allies.[67] In brief, the international order presupposed in the postwar consensus no longer existed. It was not merely that foreign developments affected domestic affairs; the line between domestic and foreign affairs itself became blurred. The conduct of foreign policy, once the province of an elite, became a more open, accessible, and disorderly process resembling domestic politics.

During this second phase of postwar science policy, regulation displaced the promotion of science as a focus of U.S. foreign policy. Earth Day at home was followed two years later by the Stockholm conference on the environment. The disappearance of the ozone layer, ocean dumping of wastes, transportation of toxic chemicals, and cleanup of oil spills became matters for international regulatory programs. Treaties sought to define standards and goals and to coordinate national efforts. The complexities of reconciling national and international laws and practices were formidable. But the efforts of national governments and international agencies to alter priorities were unremitting and at least partly successful. Multilateral development agencies modified their exclusive focus on development, paying greater attention to soil erosion, a clean environment, and appropriate technology. Development goals were, of course, never abandoned, and no resolution of the debate over the role of advanced technology for the third world occurred. But at least the "small is beautiful" argument received a full hearing and influenced the thinking of observers in the industrialized nations as well as in the third world.[68]

Spanning the years of détente, from the 1969 Nuclear Non-Proliferation Treaty to the demise of SALT II in the wake of the Soviet invasion of Afghanistan in 1979, and including such highlights as the Anti-Ballistic Missile Treaty, the Vladivostok accords, the proposed Threshold Test Ban Treaty, and the end of the Vietnam War, this period reflected the movement of national priorities away from security through reliance on technological advance to security through the control of nuclear arms. Arms control became a continuing feature of military force planning. A new emphasis on negotiation with adversaries, sufficiency rather than superiority as a guiding principle for force planning, and a more modest sense of U.S. influence in the world guided American policy.

The increasing importance of international scientific developments gave rise to an effort to formalize the place of science and technology in the machinery of foreign policy. In 1950 a committee chaired by Lloyd V. Berkner had recommended creating a unit within the Department of State to be responsible for the scientific aspects of diplomacy.[69] A small office was established that administered a program of scientific exchanges with various countries and assisted the National Science Foundation, the Department of Agriculture, and later NASA in their international programs. The office grew steadily in importance, especially after Herman Pollack, an able and energetic career foreign service officer, became its director in 1964. In 1967 Secretary of State Dean Rusk expanded its staff and the scope of its responsibilities. Later, Secretary William Rogers, in speeches and in congressional testimony early in the Nixon administration, stressed the need for departmental expertise in such matters as nuclear nonproliferation, the space (INTELSAT) treaty, and control of the seabed and the environment. He pledged that "Our basic goal is to put science and technology at the service of human—and humane—ends." [70]

In 1973 Senators Claiborne Pell and Howard Baker pushed proposals in Congress to create, respectively, the Bureau of Oceans and the Bureau of International Environmental Affairs within the State Department. These proposals were combined the following year in hearings on the Department of State appropriation for 1974 into the suggestion that a new Bureau of Oceans, International Environmental, and Scientific Affairs be established, headed by an official with the rank of assistant secretary and incorporating the existing science affairs unit.[71] The bureau was established in October 1974 to represent the increased emphasis on regulation, the environment, and the control of technology in foreign policy.

In the Carter administration, curbing the spread of nuclear technology became the overriding mission of the bureau, reflecting the president's deep personal convictions and heightened congressional concern over the dangers of nuclear proliferation. In 1978 Congress passed the Nuclear Nonproliferation Act, which added new restraints on the diffusion of civilian nuclear technology in the absence of full compliance with international safeguards. Under this law, the president was compelled to insist on "full-scope" International Atomic Energy Agency safeguards and could only authorize limited exceptions upon a specific finding of a compelling national interest. The bureau also argued for bilateral collaborative programs to support synthetic fuels and alternative energy R&D programs, in keeping with the administration's strong priorities, and pursued international initiatives to protect the environment.

The Carter administration's attempt to devise a treaty on the law of the sea, which included provisions to regulate deep sea-bed mining, was one of its most ambitious efforts. The draft of the treaty, negotiated by Special Ambassador Elliot Richardson, would have created an international entity that was authorized to issue licenses for mining in international waters and would receive part of the resulting revenues in the form of a royalty or tax. International control was justified on the grounds that the resources of the oceans were "the common heritage of mankind." This argument accorded with the popular idea of a new international economic order advanced by the nations of the Group of 77 and by third world spokesmen generally.

The proposed treaty stirred deep controversy in Congress and the administration. Its provisions seemed in conflict with the administration's efforts to deregulate the economy, the mounting public doubts about the wisdom of large government projects to develop synthetic fuels and other technological demonstrations, and the need to defend American trade and commercial interests in an increasingly competitive international environment. The controversy illustrated disagreements within the Carter administration and underscored its transitional role as national priorities began to change.

5 | The Reagan Era: A New Consensus?

THE REAGAN ADMINISTRATION accelerated four important trends already in evidence in the Ford and Carter administrations. It reasserted a strong belief in research in basic science as the central pillar of the nation's scientific enterprise (and consequently the need to restore high levels of federal support). It reflected the general disenchantment with the large demonstration projects designed to speed up the development of civilian technology and sought more reliance on traditional market mechanisms to spur innovation, though it by no means abdicated government support for applied research and the development of advanced technology. The administration also sought to deregulate the economy by eliminating strictly economic controls in key sectors, by modifying health, safety, and environmental regulations to reflect advances in scientific measurement, and by using a cost-benefit rather than a "rights" approach to environmental regulation. Finally, the administration wanted to encourage a more optimistic view of science's contribution to national defense, health, and welfare. The original postwar consensus, battered and slightly the worse for wear, seemed to have reemerged. And certainly there were strong continuities in America's postwar science policy. Despite the growing fissures among scientists during the Vietnam War, the framework of ideas, assumptions, and institutional practices that supported research never wholly broke down. Even under maximum strain, when rhetoric often reached a feverish pitch, momentum carried the system forward.

America's research system has the underlying resiliency and strength of the political system. The crisis in the relations between government and science during the 1970s, though serious enough, did not reach the point of system breakdown. But after all the din and tumult, the hand wringing and expostulation, had anything changed? Some of the nation's most thoughtful observers of science affairs found striking continuities.

Harvey Brooks, for example, noted, "There is a tendency for those close to science policy in the United States to observe changes in the system which are seen as 'revolutionary' in short-term perspective. Yet, on balance . . . what is remarkable about the post World War II research system is continuity . . . changes in the pattern of the U.S. research system have been relatively marginal." Indeed, "despite much debate . . . and cries of alarm from the scientific community, it appears . . . that the basic outlines of the 'social contract' proposed in Bush's famous report . . . have remained more or less intact, and are still broadly accepted by public and politicians." [1] The system retained its essential features and operating relationships while expanding from the modest beginnings in the 1940s to an enterprise of $62 billion in federally funded R&D in fiscal year 1988 (and a total national expenditure of twice that amount). [2]

Even more remarkable, perhaps, has been the system's capacity to accommodate changes in priorities, sudden and frequent reorientations in the focus of public attention, fluctuating public support, and heightened congressional scrutiny. Research evolved from a small-scale, executive-centered, and elite-dominated activity to a large, open enterprise. Yet it was recognizably the same system. It accommodated the military buildup of the Korean War, the post-Sputnik expansion, the energy programs of the 1970s, and the environmental movement, with the same institutions and people performing roughly the same roles. But observers were not sure what would happen when the new administration took office.

Ronald Reagan came from a state where high technology was a critical contributor to the economy. And Simon Ramo of TRW Inc., who had advised Reagan on science policy during the 1980 campaign, cautioned against the antiscience movement that he thought threatened to jeopardize the progress of American technology. [3] Still, Reagan's impulse to cut federal spending and to define clear boundaries between public and private sector functions raised fears that the new administration might cut federal support for research. Milton Friedman, University of Chicago Nobel laureate in economics and a man presumed to have influence with the new administration, stirred apprehensions when he argued that basic scientific research should not be federally supported. [4] But a more accurate indicator of what the administration's science policy would turn out to be was a Heritage Institute study celebrating the importance of scientific research to the nation's future. The report, *Agenda for Progress,* presented to the new administration shortly after the 1980 election, found a way to reconcile budget cuts in inefficient federal programs and generous treatment for research. "In a study filled with accounts of federal program failures," the report noted, "it is refreshing to find an area filled with spectacular success—space and general science."[5]

Despite some eccentric elements in the fiscal 1982 and 1983 Reagan budget requests for science funding, the directions of the administration's science policy were discernible: increased federal support for research, even though fiscal stringency was otherwise a priority.

The Reagan administration may have promised a revolution, but most of its policies stayed within familiar bounds or had their rough edges worn off in battles with Congress. When the president submitted his fiscal 1989 budget to Congress, the New Deal and most Great Society programs were intact. The Departments of Education and Energy, fleetingly considered for abolition, still stood. Despite some cuts in domestic programs, the federal budget had grown steadily: the 1989 budget request was $1.1 trillion, compared with the first full-year Reagan budget for $695 billion.[6] Rhetoric aside, was policy under President Reagan generally more evolutionary than revolutionary?

A pat yes would miss the complexities of the situation. Despite the apparent return of calm to government-science relations, and despite the aura of optimism resulting from increased R&D funding, the Reagan era reconstruction of the postwar consensus lacked some essential elements for success. The parts were reassembled, but they did not quite fit. To some degree the problems were carryovers from the era of crisis. They also reflected deep dilemmas in the relation of science to society that were sidestepped in the years of rapid expansion. Trends in the economy as well as in decisionmaking affected and complicated the task of refashioning a consensus on science policy. Unprecedented and chronic fiscal deficits—the product of recession, increased defense spending, and major tax cuts without corresponding cuts in domestic spending—in particular influenced debate on science policy throughout the Reagan presidency and set important limits on choices.

Finally, the changing nature of science itself, its tendency to require more funding in more disciplines, the revolution in computation and in the information sciences, and the shifting tactics and alliances within an increasingly fragmented scientific community added further dimensions of complexity. The result is that America now faces an unusually complex challenge: decisions of greater consequence than ever before for the U.S. research system will very likely have to be made in the 1990s. And the old formulas will yield few clear guidelines.

The Revival of Faith in Basic Research

The return to a strong national commitment to basic research can be traced through the increases in public and private funding for it, the growing strength and effectiveness of the science advisory system within the federal government, and the greater vitality of the scientific institutions and the morale of the scientists themselves. There is a complex interplay among these factors: funding affects performance, and scientific leadership influences funding levels and the directions of scientific activity.

The crisis of the 1970s in government-science relations occurred at all these levels and in the attitudes of the public toward science as well. In turn the crisis could be eased only through government actions and changes in public attitudes. Several connective threads, however, run through the web of events. A look first at important background developments will help explain how the nation under President Reagan gradually refocused its science policy.

The Presidential Science Advisory System

Scientists have considered their advice to policymakers crucial ever since the success of the wartime Office of Scientific Research and Development, so that just after the war the Research and Development Board was established in the Pentagon to continue the practice.[7] Scientific advisers were soon appointed in other cabinet departments, as well as in individual commands within the military. Some scientific leaders in the late 1940s also felt that a scientific adviser to the president would be useful, and possibly even critical, if the whole system were to work effectively.

In the summer of 1950, after the outbreak of the Korean War, President Truman asked William T. Golden, a New York investment banker who had served from 1946 to 1949 as special assistant to Lewis Strauss, chairman of the Atomic Energy Commission, and who was well acquainted with leaders of the scientific community, to study the effectiveness of the federal government's organization for scientific affairs. In particular, he was to analyze the adequacy of the way science was organized for the defense effort. When Golden presented his report on December 18, 1950, he recommended creating the position of science adviser to the president and forming an advisory committee to assist the adviser.[8]

After considerable bureaucratic infighting, Truman accepted the proposal in a watered-down version. In April 1951 he named Oliver Buckley, recently retired as president of AT&T's Bell Laboratories, as science adviser to the director of the Office of Defense Mobilization, a unit of the Executive Office responsible for industrial mobilization and the administration of wartime controls. Buckley was also appointed chairman of a part-time advisory committee to be part of ODM. But Truman specified in his letter of appointment that Buckley would also "from time to time" give advice to the president.[9]

Buckley was not aggressive. He thought that advisers should respond to requests for advice and not seek to impose their views on decision-makers. Further, the committee was to avoid budgetary matters and was not to interfere with the responsibilities of existing agencies. It was to work through the departments and agencies and give long-range advice on matters that cut across bureaucratic jurisdictions. Opinions differ on how effective the committee was in its first year of operation, but clearly it did not aggressively seek to influence high-level executive branch politics.

In the wake of Buckley's resignation in June 1952 because of ill health, members of the committee decided that the functions they performed were needed but that their location in ODM did not offer the right administrative channel. They recommended to Truman that they report directly to him or to the National Security Council, which could deal more effectively with long-range problems. When the president did not accept this recommendation, the committee remained in ODM with Lee A. Dubridge serving as part-time chairman and adviser to the director of ODM. Discouraged, the members considered resigning and recommending that the committee be abolished but decided to await the results of the 1952 election and to review their role with the new president.

President Eisenhower met with the committee in the spring of 1953. When asked whether there were issues on which committee advice could be of use, he replied, "Surprise attack. I worry about whether one morning we might not wake up, whether the nation could be destroyed in a surprise atomic attack."[10] Energized by his interest, the committee subsequently produced confidential reports recommending the development of intercontinental ballistic missiles, outlining steps to improve the alert status of U.S. strategic bombers, and exploring technical problems of air defense.

These matters were neither narrowly technical nor matters of science as such; rather, they were broad policy concerns with technical dimensions. The term science adviser, though sanctioned by popular usage, is thus misleading. When they have been effective participants in making policy, these advisers have not been didactic pedagogues merely talking about science but have participated on political terms defined by the policymakers.

Oliver Buckley's conception of his duties had not been wholly misguided. Except for Herbert Hoover and Jimmy Carter, both trained as engineers, twentieth century presidents have shown little knowledge of science or curiosity about it. Nor have they been particularly concerned about scientific institutions. Neither Truman nor Eisenhower was much interested in efforts to strengthen scientific institutions (in his second term Eisenhower's perspective broadened). Although each made important decisions drawing on scientific advice, especially for defense and space policy, they assumed that while the health of science, like the health of the economy, was important to the nation, their responsibilities only rarely required explicit attention to the health of science as such. The well-being of science was best left to the mission agencies, which nurtured it for their own purposes, to the National Science Foundation (then so small as hardly to stir the attention of policymakers), and, of course, to the universities, industrial laboratories, nonprofit institutions, and the scientists themselves. Both Truman and Eisenhower were also concerned about the effect of government expenditures on the economy and sought to keep spending, including funds for R&D, under control.[11]

A notable exception to such inattention was Eisenhower's decision to support U.S. participation in the 1957 International Geophysical Year. In 1956 a small group of scientists, led by geophysicist Lawrence Gould, paid a call on Assistant Budget Director William D. Carey in the old Executive Office Building. Gould explained the advantages of U.S. participation in the IGY and estimated the cost. Carey told the group to wait while he went downstairs to brief Budget Director Percival Brundage. He suggested to Brundage that the president might wish to give the matter priority. Brundage told Carey to wait, walked to the White House, and returned in ten minutes to say that the president enthusiastically supported the idea. The elated scientists left with a $10 million commitment for U.S. participation.[12]

The Soviet Sputnik launches in the fall and winter of 1957 dramatically broadened the role of the presidential scientific advisers and drew the president more closely into the institutional issues of science policy. Eisenhower created the post of special assistant to the president for science and technology and, on the advice of his friend, I. I. Rabi of Columbia University, named James Killian of MIT to the position.[13] Killian, an

academic administrator who enjoyed the confidence of the scientific community to an unusual degree, was very effective in planning for the post-
Sputnik expansion of American science and in advising the president on
policy matters with technical dimensions. He had had wide experience
in defense and space matters, had participated in planning U.S. missile
programs, and had served with distinction on scientific panels in the early
1950s. Eisenhower also elevated the status of the advisory committee in
the ODM to that of Presidential Science Advisory Committee (PSAC).
The group elected Killian chairman.

Killian used the several duties of his position—adviser to the president, source of expertise for budget officials and other presidential advisers, convenor of the advisory committee, and interagency facilitator and
coordinator—to help produce coherent policy in a period of extraordinary expansion of government support for R&D and education. The
NSF's budget alone increased from $50 million to $133 million between
fiscal 1958 and 1959, and funds for science education also increased
sharply.[14] The National Aeronautics and Space Act of 1958 created
NASA, with the laboratories and organization of the National Advisory
Committee on Aeronautics as its nucleus and a greatly expanded budget.
The National Defense Education Act of 1958 became the prototype of
other steps to expand the federal government's support for higher education apart from funding for specific research projects.[15]

While the science adviser and PSAC dealt with the expansion of the
research system, their most important functions were presidential advice
and policy oversight and coordination.[16] They strengthened the president's hand, and to this end fell into a natural alliance with budget officials, which helped check the power of the large agencies, or at least
ensured that presidential priorities were reflected in space and defense
policies. They cooperated less closely (and occasionally fought) with
budget officials about R&D budgets of smaller agencies because they
often tried to foster agency technical activities while the budget officials
tended to view such efforts as comparable to public works projects and
other pork barrel programs.

The tensions of the Vietnam War brought increasing strains that
tended to limit access to the president. The major difficulty lay in reconciling the habits of scientists, long accustomed to open communication
among colleagues, with the discipline, discretion, and confidentiality expected of important presidential advisers. The science adviser increasingly faced the choice of distancing himself from colleagues, who in-

cluded critics of the war effort, or risking exclusion from presidential decisionmaking.

Even before the escalation of the war and the antiwar movement, however, conflicts among the several functions of the science adviser began to emerge. The scientific advisers and their allies in the Bureau of the Budget suffered a setback in 1961 when an amendment to the National Aeronautics and Space Act revised the membership and functions of the National Aeronautics and Space Council by bringing it within the Executive Office of the President. The vice president was installed as chairman. This action divided staff responsibilities for planning space policy, but it was defended on the grounds that the complex operations of the mushrooming space effort required a special staff unit. There was some logic to the argument. The science adviser could not hope to be the only source of advice to the president on important technical matters. Nor would he want to become too heavily involved with any one program. But the creation within the Executive Office of the President of such specialized units reflecting powerful agency and constituency interests inevitably complicated the tasks of the adviser and his budgetary allies, who saw themselves as responsible for integrating policies.

On June 14, 1961, the Subcommittee on National Policy Machinery of the U.S. Senate Committee on Government Operations issued "Science Organization and the President's Office," a report recommending the creation of a permanent office of science and technology to assist the president.[17] Concerned over the fragility of the office, President Kennedy's science adviser, Jerome B. Wiesner, persuaded Kennedy to issue Reorganization Plan no. 2 of June 1962 formally establishing it.

Some experienced observers cautioned against the step. They feared that the science adviser would be too deeply involved with the minutiae of managing an office and seeking appropriations from Congress and that consequently the advisory system might become less flexible, less responsive to the president, and less concerned with mobilizing the best outside talent for part-time service to the nation. Experience would show that these fears were justified.

The period from the appointment of James Killian to the assassination of President Kennedy has often been considered the high point of the advisers' influence in the White House, in part because the issues confronting them were relatively circumscribed and highly technical.[18] But the job became increasingly difficult as the task of balancing competing functions became more demanding. As the unity of the scientific community dissolved, as the science adviser was called upon to deal with complex social issues, and as immediate post-Sputnik confidence gave way to more critical public attitudes toward science, the firm mandate that had made the appointment of a presidential adviser seem such a good idea faded. For a time inertia contributed to the continuance of the office: Presidents Johnson and Nixon preferred not to offend a constituency of some importance. But presidents and their intimate advisers have never believed there must be a science adviser, nor have they believed strongly in a committee of distinguished outside scientists located in the Executive Office. This belief has been an article of faith among scientists.

Thus it should not have come as a total surprise when, after growing controversy about public statements critical of administration policies by members of the PSAC, President Nixon decided to accept the pro forma resignations of the members at the end of his first term and to appoint no new ones. On January 3, 1973, the White House announced that Edward E. David, Jr., had resigned as science adviser.[19] Some of the adviser's duties were then transferred to H. Guyford Stever, director of the National Science Foundation.

Abolishing the office did not, of course, eliminate scientific advice to the president. Like his predecessors, Nixon sought technical consultation when he felt, or was persuaded to feel, that he needed help. Mch has been written about the dire consequences of the absence of a science advisory office in full panoply in the White House, the commentary often preoccupied with formal structure and reporting channels.[20] But it is value conflicts and disputes about the proper ends of government that are principally at issue in questions of White House staff organization, not neutral principles of management. Convictions about policy, not the presence or absence of a science adviser or how frequently the adviser sees the president, account for major presidential decisions.

The absence of a White House science adviser may, however, have compromised the coherence of some policies. Science advice now came to the president from the mission agencies and other interested parties. The presence of a science adviser and a strong technical staff in the White House might have improved the quality of staff work on technical matters and aided the OMB, which by itself was a less formidable counterweight to the large mission agencies than the combination of OMB, OST, and PSAC. This absence could have meant that the White House sometimes reacted to a crisis hastily and with less attention to harmonizing technical policy conflicts and discordant program elements.

For instance, soon after he abolished the science advisory system, Nixon was forced by the energy crisis of 1973–74 to name a new adviser, John A. Love, on energy policy issues and to create the Energy Research and Development Council, consisting of experts from industry and the universities to support the adviser. Confusion ensued as the government sought to come to grips with the crisis. In December 1973 Atomic Energy Commission Chairman Dixie Lee Ray presented a plan for a five-year, $10 billion energy R&D program, and recommended the creation of an Energy Research and Development Administration no later than July 1, 1974.[21] Once implemented, the energy research and development program included large demonstration projects considerably more elaborate than research projects, and they were not always linked either to ongoing research efforts or to the needs of the marketplace. A totally congruent energy policy was simply not likely as the nation struggled with the crises of 1973–74 and 1978–79. But some of the anomalies in the ensuing research policies and in their relationship with wider policies might have been avoided with an OMB-OST-PSAC system in the Executive Office.

At any rate, concern over the consequences of a vacuum in scientific advice at the center of government led to congressional efforts to reestablish the advisory system soon after President Nixon abolished it. Ellis Mottur, an aide on science affairs to Senator Edward M. Kennedy, was instrumental in having a provision for recreating the White House science advisory system added to S.32, the National Science Policy and Priorities Act, which linked federal R&D spending to a fixed percentage of GNP growth and established parity between defense and other federal spending for research.[22] The administration strongly opposed the measure because it would have created a semi-independent agency, the Civil Science Services Administration within the NSF, authorized to spend $800 million on civilian technology development over three years. The administration also objected to the effort to impose congressionally mandated staffing arrangements on the president. S.32 failed as the energy crisis overwhelmed other aspects of civilian technology policy, but the thinking behind the bill continued to have broad support in Congress.

A second source of concern about the demise of the presidential science advisory system was the Council of the National Academy of Sciences. In February 1974 the council appointed James R. Killian, Jr., chairman of a committee to study the matter of providing scientific advice to the president and to the management of federal R&D programs.

A third group, including physicist Edward Teller and attorney Oscar Ruebhausen (who had been OSRD general counsel and had initially discussed with Vannevar Bush the proposal by Oscar Cox for a report to Roosevelt on postwar science), gathered around former Governor Nelson Rockefeller of New York. Rockefeller was devoting his energies to the Commission on Critical Choices for Americans, an ambitious study group he had created and partly funded to look into the longer-term problems facing the nation. He had become convinced of the need for a White House science adviser, and when he was appointed vice president in August 1974, he persuaded President Ford of the need for the post. In December 1974 the president formally requested him to study the question and to make appropriate recommendations.[23]

Through Rockefeller's efforts and those of Senator Kennedy and others in Congress, President Ford was able to sign the National Science and Technology Policy, Organization, and Priorities Act in May 1976. The act established a new Office of Science and Technology Policy (OSTP) within the Executive Office of the President, replaced the old interagency council with a new Federal Coordinating Council for Science, Engineering, and Technology to be chaired by the OSTP director, created the Intergovernmental Advisory Panel to bolster the technological capabilities of the state governments, and mandated a study of the federal government's R&D effort by a temporary advisory committee. The act did not reestablish a permanent science advisory committee, principally because the Freedom of Information Act required open meetings. NSF Director Stever was sworn in on August 12, 1976, as the science adviser.

With the election of Jimmy Carter, the future of the office was in doubt. Carter was committed to streamlining the federal government, including the Executive Office. The OSTP did come close to being transferred out of the White House and having its staff positions sharply cut, but Carter was persuaded to retain it, and Frank Press, a distinguished geophysicist from MIT, was formally sworn in as adviser on June 1, 1977.

Press pursued policies that anticipated the directions of the Reagan science policies. He acquainted senior OMB officials with trends in academic science, which helped pave the way for increased support for basic research and eased tensions between university administrators and government officials over federal reimbursements of indirect costs. He worked closely with OMB officials to reduce federal outlays for large energy demonstration projects that showed little promise of early application. And he also backed the review of civil technology, which directed attention toward antitrust and patent policies, regulation, adjustment assistance, and other nontechnical factors affecting the climate for innovation.[24] His office, with the aid of Gilbert S. Omenn, a physician and OSTP deputy director, sought to upgrade the quality of scientific research supporting the regulatory process for occupational health, safety, and the environment. Finally, increased spending on defense research, in the context of heightened awareness of defense needs, began during the Carter administration.

Developments in Industry and the Universities

While observers worried about scientific advice at the White House, concerns also grew about the conditions in the laboratories where the research was being done. In the 1970s the universities had undergone student disturbances, political tensions, austere budgets, and increasing regulatory burdens. Inflation was severe. Capital facilities, laboratory instrumentation, and the general infrastructure supporting research had endured a long period of deferred maintenance and relative federal neglect. But in 1977, *The State of Academic Science,* a study by the National Science Foundation and the Association of American Universities, detected a shift in the national mood. After a decade of attacks from the Right and Left for being hotbeds of radicalism or tools of imperialist military power or sources of disengaged elitism, universities now seemed to receive a more sympathetic hearing for their problems from both policymakers and public.[25]

The study concluded, "When federal research support grew rapidly from the late 1950s until the mid-1960s, science departments built a base of equipment that helped carry forward the research effort. However, the equipment . . . has now begun to age and wear out in many departments . . . [and] the long-range consequences of a deteriorating equipment base for American science are serious." [26] In 1980 the National Science Foundation's *Five Year Outlook* warned that the research system could no longer operate effectively on the basis of earlier capital investments. "The federal government's decision to reduce allocations for capital investment and equipment was a rational response to budgeting problems. However, the period of low investment was so protracted that many research installations have become, or are becoming, obsolescent. Some experiments simply cannot be performed in existing facilities with an earlier generation of equipment." [27]

On the positive side, the campus disturbances had gone, and the fears that internal reforms induced by wartime protests would cripple university governance proved largely groundless. The universities adapted to affirmative action requirements and other regulatory measures with fewer difficulties than expected. They were also rebuilding their relationships with private industry, which in some cases had atrophied during the period of rapid expansion of federal research support.

Meanwhile, industry was also changing. R&D activity had begun to pick up again. Trends toward a "postindustrial" economy were becoming evident in the rapid expansion of the nonmanufacturing or service sector, which accounted for much of the increase in science and engineering employment, and in the rising employment of scientists, engineers, and technicians in an otherwise static employment market in the manufacturing sector, reflecting an increasingly high-technology work force.[28]

Employment trends for scientists and engineers parallel the trends in industry R&D expenditures. Employment of scientists and engineers engaged in industrial R&D declined as a percent of the total U.S. labor force, dropping some 16 percent from a peak in 1963 to a low in 1975, presumably reflecting the reduced demand caused by falling federal research expenditures in space and defense. After 1977, growth in self-financed industrial research began to pick up the slack. Technical employment began to increase much faster than general employment in both nonmanufacturing and manufacturing companies. From 1977 to 1980 the annual growth in employment of scientists, engineers, and technicians in manufacturing was 6 percent, compared with only 1 percent for the growth of the manufacturing labor force as a whole. Science and engineering employment as a fraction of total employment actually grew most in those industries experiencing a decline in overall employment. Thus in some thirteen industries averaging a 3 percent decline in employment, scientific and technical employment increased by 11 percent.[29]

Scientists and engineers as a part of the total U.S. work force rose from 2.4 percent in 1976 to 2.6 percent in 1980, 2.9 percent in 1982, 3.3 percent in 1984, and 3.6 percent in 1986. This reflected increases in federal R&D expenditures beginning at the end of the Carter administration and the general demand for higher skills in the work force. Employment in science and engineering substantially exceeded annual growth rates in total U.S. employment and GNP (table 5-1).

The expansion in technical employment and industrial R&D was driven by challenges from Japan, the newly industrializing economies of Asia and Latin America, and a resurgent Europe. Because the Japanese conquest of market segments clearly rested in part on manufacturing skills, some traditional low-prestige skills such as industrial and mechan-

Table 5-1. **Average Annual Increases in Employment in Science and Engineering, Selected Periods, 1976–86**
Percent

Category	1976–80	1980–86	1976–86
Scientists and engineers	4.6	7.9	6.6
Scientists	5.2	7.8	6.7
Engineers	4.3	7.9	6.4
U.S. employment	2.8	1.7	2.1
Real gross national product	3.0	2.4	2.6

Source: National Science Board, *Science and Engineering Indicators, 1987* (Washington: National Science Foundation, 1987), table 3.3.

ical engineering witnessed a boom in employment as U.S. companies sought to upgrade their manufacturing processes.[30] As a result, the American Society of Mechanical Engineers, after intensive lobbying, persuaded the NSF to launch a new program to strengthen research and training in mechanical engineering.[31]

The application of the computer to the manufacturing process eventually meant a closer integration between design, manufacture, and quality assurance, and a breakdown of some traditional status distinctions in engineering. U.S. companies making machine tools, consumer products, and heavy manufactures rushed to automate, to retrain workers and involve employees in problem-solving exercises, and to institute statistical process controls. Even small companies could improve quality by adopting computer-aided design, although the scale of their operations often could not justify computer-aided manufacturing. Aerospace, telecommunications, biotechnology, and pharmaceutical companies, already research- and technology-intensive, increased still further the educational levels of their workers and the number of engineers and scientists. Spectacular progress occurred in microelectronics and computing as the number of bits of information that could be processed on sophisticated integrated circuits and microprocessors grew exponentially.[32]

Advances in molecular biology at leading academic centers in Boston, San Francisco, and San Diego, which spurred the formation of new genetic engineering and related companies in the 1970s, illustrated the importance of close links between industry and universities. The short time span between discovery and potential application tended to confirm the premise that funding basic research could provide the conceptual breakthroughs, the human resources, and the general infrastructure for a flourishing industry. Beginning in the 1970s, the phenomenon of the research park, as exemplified by Route 128 near Boston, the Silicon Valley near

Stanford University, and the Research Triangle near Raleigh-Durham, North Carolina, provided further confirmation that incentives toward collaboration were important for both industry and academia.[33] Collaboration was sometimes aided by government action, but more often took place without explicit state or federal involvement. Gifts of equipment, fellowship support, and unrestricted contributions from industry to universities increased. Patent reforms spurred further university-industry cooperation in the 1980s.[34]

Perhaps one of the most striking developments was the turnaround in the auto industry's attitude toward R&D. Faced with dramatic Japanese penetration of U.S. markets in the 1970s, it has since transformed itself from a mature industry to one relying heavily on advanced technology both in product design and manufacturing processes. The commitment to technology has not been an unqualified success, however. General Motors, which invested most heavily in automation, found itself plagued by difficulties in its most technology-intensive plants and later was saddled with high initial investment costs and overcapacity. Ford Motor Company, which proceeded more cautiously with automation and emphasized quality, surpassed GM in profitability in 1986 and in 1987 registered net earnings of $4.6 billion. Aggressive competitiveness from foreign manufacturers continued, however, with the Japanese automakers, notably Toyota, and also Honda among the "transplants" in the United States, continuing to show shorter product cycles and higher productivity.[35]

Finding the right combination of technology investment, management strategy, labor-management cooperation, cost and quality control, and productivity increase remains a continuing challenge to the U.S. auto industry. There is no safe plateau for any one company or for the industry as a whole. The auto industry's recovery was aided to an undetermined degree by the voluntary restraint agreement with Japan that limited the Japanese to exports of 1.8 million passenger vehicles to the United States, but the VRA allowed and encouraged the Japanese to export bigger, more expensive autos, thus attacking market segments that the U.S. auto industry had hitherto regarded as secure.

A Turnaround in Government R&D Funding

Along with reestablishment of a science adviser in the White House, increased interest in the state of university research, and industry's renewed emphasis on R&D, the new mood toward science could also be seen in federal research priorities. Federal funding for basic research in universities, measured in real terms, had declined by 9 percent from 1968 to 1976.[36] President Ford's fiscal 1977 budget included $50 million additional funding for basic research; Congress eventually approved $22 million.[37] Ford's fiscal 1978 budget request included an 11 percent increase in budget authority for R&D overall, and his budget message singled out basic research as one of two areas needing special attention; advanced defense systems was the second, as the Nixon-era détente began to be replaced by a gradual awareness of the need to rebuild the nation's defenses.

The Carter administration did not make major changes in Ford's budget requests in its February 1977 amendments, but it did cut military R&D by $200 million, an amount partially restored several months later when the B-1 bomber was canceled, and altered some energy priorities. Carter's first complete budget request, for fiscal 1979, emphasized basic research in the mission agencies, which had, in their emphasis on more targeted research, neglected basic research during the previous decade. Budget Director Bert Lance sent a letter early in the budget review process to the heads of these agencies urging them not to slash basic research, and Science Adviser Frank Press reinforced this message in meetings with agency officials.[38]

President Carter's fiscal 1980 budget reinforced the previous year's emphases despite the mounting inflationary pressures and public concern about high government expenditures that were leading the administration toward greater fiscal restraint. Carter also refused to back away from his commitment to NATO for a 3 percent real increase in U.S. defense spending. Thus the budget moved toward favorable treatment of R&D, especially basic research, and increased defense spending even while striving to reduce overall government expenditures.[39]

In the last week of March 1979, when the nation's attention was focused on the signing of the historic Israeli-Egyptian peace treaty, Carter sent a special presidential message on science and technology to Congress—the most comprehensive presidential statement of federal policies toward R&D and the first since Nixon's 1972 message. It was also the first major government report for some time to echo the Bush report's faith in basic research as the foundation for achieving important government missions. The 7,500-word message's purpose was to articulate "a science and technology policy for the future. . . . The thesis is that new technologies can aid in the solution of many of the Nation's problems. These technologies in turn depend upon a fund of knowledge derived from basic research. The federal government should therefore increase its support both for basic research and, where appropriate, for the application of new technologies." [40]

The message did not produce a new blueprint for science and technology; it served mainly as defense of the budget and as a conciliatory gesture toward Congress (some members were irked at the failure of the executive branch to comply fully with the reporting requirements of the 1976 act reestablishing the presidential science adviser). But the message did articulate the broad directions of policy. Besides reaffirming the importance of basic research, it noted the need to stimulate technology development in the private sector. But government was to stay clear of R&D programs with near-term commercial payoffs. These were better left to the workings of the marketplace, although the private sector's capacity to innovate was one of the great challenges facing the nation.

The message spelled out national security concerns in some detail, stressing the importance of maintaining technological leadership in weapons, using technology to reduce the costs of expensive defense systems, building a strong research base to provide for future defense needs, preventing the export of sensitive technologies, and using advanced technology for verification in the pursuit of arms limitation agreements.

When the fiscal 1981 budget was being prepared, the administration's main priorities had been to combat inflation by controlling expenditures, to secure the Senate's ratification of the Salt II Treaty, and to press for enactment of a windfall profits tax on oil. The budget was tight and projected the lowest federal deficit in seven years. Nonetheless, basic research was exempted from the general policy of fiscal constraint until a sharp increase in the consumer price index prompted last-minute revisions. Expenditures were sharply cut back. R&D spending still suffered less than most other relatively "controllable" expenditures, although it fell below a real (constant dollar) increase. [41] The Soviet invasion of Afghanistan heightened the importance of defense as a national priority, most notably seen in Carter's lame duck fiscal 1982 request, which showed dramatic increases for defense and defense R&D. [42] Events abroad thus reinforced the trends at home and seemed to propel the nation toward greater reliance on a strong base of science and technology. In this context the Reagan term of office began.

The Reagan Administration's Science and Technology Policy

Like most broad impulses in American politics, the Reagan revolution was "a tendency rather than a theory, a collection of precepts that fit together uneasily, and an operational code driven by a new mix of political necessity and ideological orthodoxy." [43] The first expressions of policy for science and technology followed the main lines of President Reagan's macroeconomic policies and were shaped by officials from the OMB, including Director David A. Stockman. In the Reagan revisions to the Carter fiscal 1982 budget, funding for civilian R&D was scaled back from $43.9 billion to $40.6 billion, but defense R&D was increased by $1.5 billion (7 percent more than Carter requested, a figure that already represented a 23 percent increase from the previous year's request).[44] Nondefense basic research was spared from heavy cuts, at least in the physical sciences and engineering. This commitment to support basic research was to grow remarkably during the rest of the Reagan presidency.

In the fiscal 1982 budget revision, however, protection for basic research in the physical sciences and engineering was accompanied by an attempt to cut federal funding for social science research in half, eliminate the NSF's science education program altogether, and reduce spending for the biomedical sciences substantially. Social science was the bête noire of Stockman, who had assailed its alleged ideological bias six years earlier.[45] NSF's science education program had come under fire from conservatives in both parties because it allegedly promoted a liberal, antireligious (or "secular humanist") ideology in the nation's public schools.[46] Congress, however, rebuffed these administration proposals.

Within the next several years the administration abandoned its opposition to the social sciences and became an enthusiastic supporter of science education programs. It continued efforts to slow the growth of spending for biomedical research, but these increasingly became mere ritual as Congress invariably restored proposed cuts and even increased the NIH budget each year. In Reagan's second term an OMB proposal surfaced to "privatize" the intramural research function of the NIH (the grantmaking function was to remain untouched). In effect, the idea was to create a new private university off budget that would compete for research funds from the government and industry on a par with other private universities.[47] This proposal encountered stiff opposition from Congress and the academic community, and quickly dropped from serious consideration.

The Science Advisory System

The early wavering in the administration's science policies stemmed in part from the lack of appropriate knowledge and experience. The administration tried to recruit a senior scientist from industry as science adviser, but initial efforts were unsuccessful. A stumbling block was that the science adviser, in the scheme of White House organization, was to report to Domestic Policy Adviser Edwin Meese rather than directly to the president.[48] To some scientists this appeared to diminish the status of the position, although the president's top advisers considered the proposed arrangement no different from that of the national security adviser, who would also report to the president through the domestic policy adviser. Was this a case of scientists attempting to dictate presidential staff arrangements, or would a science adviser so far removed from the councils of power be in fact unable to serve the president well? Whatever the merits of the opposing views, there was no adviser on hand during the early months of the Reagan administration when many of the major policies were set.

This absence did not affect priorities (any science adviser appointed by the president would have embraced the administration's goals). Nor did it leave funding for research without advocates. Strong belief in basic research (at least in the so-called hard sciences), in more spending for defense R&D, and in steps to strengthen technology to improve the competitiveness of American industry were important parts of the administration's policy from the start. The science adviser's absence did mean, however, a certain incoherence in policy, of which the administration's attacks on social science and on science education, at the same time that it was seeking a more scientifically literate society, were a symptom. Science received generous new support, but the various policy emphases were not effectively integrated.

In August 1981 the arrival of George A. Keyworth, a physicist and defense specialist from the Los Alamos National Laboratory, as science adviser partly met the need for more coherent direction in science policy. Relatively inexperienced in Washington administrative politics, Keyworth started slowly but soon became an effective spokesman for the administration. In February 1982 he established a White House science council, chaired by Solomon J. Buchsbaum of AT&T Bell Laboratories, to advise on issues affecting science agencies and to study programs and policies that cut across agency responsibilities.[49] But Keyworth never had sufficient resources to build a strong staff and relied mainly on officials borrowed from federal agencies. His staff experienced rapid turnover and a resulting lack of continuity.

Critics complained that by reporting to Edwin Meese rather than to the president, the science adviser did not have sufficient weight or stature to influence policy or command respect from the government agencies. The White House Council, some critics believed, did not tap a wide enough pool of scientific talent and did not spend enough time on its task (it typically met two days a month). Finally, critics charged that the OSTP staff was weak and ineffective. For his part, Keyworth retaliated that the critics did not understand the need for the science adviser to fit the president's style of operation and to be loyal to the administration.[50] Buchsbaum, a veteran of many advisory positions, including membership on the President's Science Advisory Committee during the Nixon administration, observed that "critics expect miracles these days [and] such miracles are harder to come by. After World War II and even after the shock caused by Sputnik, America's technological, military, and economic power was of such magnitude that we could set our own and even the world's agenda. And stick pretty much to it. The world has caught up with us. We now have to react to the agendas set by others." Disputing those who considered the White House Science Council ineffective "presumably because it lacks the word 'Presidential' in its title," Buchsbaum noted that appearances did not always conform to reality. "During the Nixon days, for example, the PSAC was still held in high esteem by the scientific/technical community. Unfortunately, neither the power stucture within the White House nor, indeed, the various departments and agencies of government shared this attitude. As a result, the committee was largely ineffective."[51]

A comparison of the science and the national security advisory systems is perhaps instructive. Reagan's first national security adviser, Richard Allen, did not report directly to the president; partly as a result, one may argue, national security affairs received inadequate high-level attention. And Allen was, at any rate, unable to exercise the traditional role of honest broker effectively and failed to resolve disputes between the State and Defense Departments. His successor, William Clark, had easy access to the president but lacked experience in foreign affairs and national security policy. The Iran-Contra crisis of November 1986 revealed the flaws in the system, and in the wake of the Tower Commission report and a congressional investigation, the staff was reorganized. After that the NSC performed effectively.

In contrast, the Office of Science and Technology Policy has never contributed to a crisis or had an obvious breakdown of function. It has been a minor player in high-level executive politics. The effects of a weak science advisory system are difficult to document; unlike the NSC system, there are no life-and-death foreign policy issues that seize national attention. But the absence of central direction in resolving small policy disputes, the failure to set priorities and integrate agency policies, and the erosion of quality control over the government's technical programs do have consequences over time. If the OSTP is weak, agency enthusiasms may not be carefully evaluated, critical breakdowns may go undetected, and matters that cut across jurisdictions may be inadequately addressed. The *Challenger* shuttle disaster, while not directly traceable to the OSTP system, resulted from a steady deterioration in the quality of an agency's technical operations. A more vigorous OSTP might have picked up signs of trouble earlier through its contacts with NASA's technical staff.

The rapid buildup in defense R&D early in the Reagan administration made more project funds available to the nation's universities, but there was little consideration of needs for plant and research equipment. This problem began to be addressed in December 1983 when the DOD-University Forum was created. The forum allowed university and defense officials to exchange views about the administration's increased security restrictions on scientific communications, the need to upgrade aging laboratories, support for indirect costs, and other sensitive issues in DOD's relations with universities.

George Keyworth resigned as science adviser as of December 31, 1985. His deputy, John P. McTague, was named acting director of the OSTP on January 1, 1986, but did not want to serve beyond an interim period. He left on May 23, 1986, and was succeeded by Richard G. Johnston, who served as acting director until October 2, 1986, when William Graham was confirmed as science adviser and OSTP director. Graham served through the end of Reagan's term and into that of George Bush, when he was succeeded by Yale physicist D. Allan Bromley. The position was upgraded from special assistant to assistant to the president for science and technology.

National Defense and Basic Research as Budget Priorities

The Reagan administration did not state a policy for science and technology in its initial pronouncements; these concerns were not high on its agenda. But its science policy can be deduced from the budget decisions made by the president and his advisers in the first three months. First, overall government expenditures were to be cut beyond what had already been projected in the tight Carter request for fiscal 1982. Second, heavy cuts in nondefense spending would allow additional defense increases beyond the projected Carter increase. R&D expenditures were largely a reflection of these broad objectives. As usual, there would be no R&D budget as such: the aggregate spending for research would be the total of each agency's requests based on their own assessments of technical needs. The administration's philosophy would, however, guide agency choices. Thus development efforts were normally to be supported only where the government itself was the customer (as in most defense systems and many space systems). Government R&D was not to be undertaken to aid the transition to the marketplace of technologies with near-term commercial potential.

The revised budget request for 1982 included the following measures:

—Defense R&D, reflecting the overall increase in the DOD budget request, was $1.5 billion (or 6 percent) more than Carter's request, most of the funds being slated for development.

—Nondefense R&D was to be cut $3.3 billion, with the Carter administration's energy program a major target ($1.5 billion from the Department of Energy budget) along with other commercial demonstration and technology-assistance programs.

—Tax reductions, not federal programs, were to provide incentives for technological innovation in industry.

—Federal hiring, travel, and use of consultants were to be restricted.[52]

The revisions sought a total reduction of $44 billion (or 6 percent) from the Carter request. Despite its zeal in cutting nondefense expenditures, its aversion for the Carter administration's commercial demonstration projects, and its general interest in narrowing the areas in which the federal government should act, the Reagan administration did not attack the concept of federal support for basic research. Some cuts were requested but at a very modest level ($5.3 billion overall as compared with Carter's request of $5.4 billion—and still a slight increase over the actual fiscal 1981 appropriation).[53] The Reagan budget cut basic research, applied research, and development 4.5 percent as against 6 percent for the rest of government expenditures. Some new initiatives, such as the NSF program of support for improvement of scientific instrumentation in university laboratories, were postponed.

The closest thing to an early policy statement on R&D is a passage from a February 1981 OMB document, *A Program for Economic Recov-*

ery, which generally endorsed the value of research but did not exempt it from suffering budget cuts: "The merit of research and development is without question. However, in times of fiscal austerity even some promising investments in science and technology must be restrained and new undertakings postponed." [54]

The administration began the most active phase in its science policy by confirming Keyworth as science adviser in August 1981 and by filling the major subcabinet posts for science and technology in the departments.[55] Keyworth began to grope toward a strategy that would reconcile aggressive advocacy for funding basic research with the administration's desire to curb spending. In a speech before the AAAS R&D Colloquium in December 1981 he seemed to suggest a more selective approach to the support of basic research. The United States could not hope to be the world leader in all fields of science and should thus embrace excellence as a principle to guide federal policies so as to ensure adequate funding in key areas. He opposed such "magic formulas" as 3 percent real growth a year for basic research or 5,000 new and competing grants at NIH. It was unclear what criteria he would use to determine excellence, but he hinted that the disciplines most deserving would correspond closely to those most likely to produce promising applications. Keyworth backed off, however, from this policy of selective excellence in succeeding months.

In the course of preparing the fiscal 1983 budget (and the "Special Analysis K" summarizing the administration's R&D policies) and the *Annual Science and Technology Report* sent to Congress by Keyworth in April 1982, the administration assumed an attitude toward basic research more reminiscent of the Carter and Ford administrations.[56] As stated in "Special Analysis K," the federal government would support R&D activities in response to the needs of a specific government agency or to a broader national need. While the administration's 1983 budget sought to reflect "a clearer distinction than has been the case in the past between the responsibilities of the Federal Government and those of the private sector," there was a broad responsibility to support "basic research across all scientific disciplines." Support of technology development, other than for specific government missions, would be limited to those "technologies requiring a long period of initial development, such as fusion power," for which the risks were too high for the private marketplace to bear and the potential long-term payoff to the nation was large. The main federal responsibility remained to "provide a climate for technological innovation which encourages private sector R&D investment." Thus "the administration's R&D policy is part of its overall economic policy," and the challenge was to bring together policies on research spending, regulatory relief, and taxes to achieve administration goals.[57]

There were, however, traces of the idea of greater selectivity in the approach to the 1983 budget. High-energy physics, for example, was to get a bigger increase than other physical sciences. Space astrophysics also got a significant increase, while funds for planetary astronomy were cut back. Some of the proposed cuts and increases appeared to reflect eccentric rather than coherent criteria of choice. NIH biological research programs, for example, were constrained beyond what one might have expected in light of the administration's stated objective to support basic research in all disciplines, while funding for NSF's physical sciences programs was especially generous. Most of these anomalies were smoothed out by Congress. But the effect was to promote a certain degree of anxiety and confusion among researchers. Thus even though the nation was embarking on a period of spectacular growth in R&D spending, tensions remained between the government and the research universities.

By fall 1982, when the fiscal 1984 budget was being prepared, the administration's commitment to increased spending for R&D (and the importance of science and technology to its policy goals) was clear. The twin priorities were national defense and basic research. The 1984 budget continued the defense buildup and sought increases above the rate of inflation for basic research. The nation's security would be greatly aided by the fruits of advanced technology, and in the long run, support for basic research would strengthen the economy and produce jobs. The administration's faith in science and technology is perhaps most eloquently spelled out in President Reagan's January 23, 1983, State of the Union address:

Education, training, and retraining are fundamental to our success as are research and development and productivity. . . . We Americans are still the technological leaders in most fields. We must keep that edge, and to do so we need to begin renewing the basics. . . . To many of us now, computers, silicon chips, data processing, cybernetics, and all the other innovations of the dawning high technology age are as mystifying as the workings of the combustion engine must have been when that first Model T rattled down Main Street, U.S.A. But as surely as

America's pioneer spirit made us the industrial giant of the 20th century, the same pioneer spirit today is opening up on another vast front of opportunity, the frontier of high technology.

In conquering the frontier we cannot write off our traditional industries, but we must develop the skills and industries that will make us a pioneer of tomorrow. This administration is committed to keeping America the technological leader of the world now and into the 21st century.[58]

President Reagan's use of the promise of science and technology as a dominant motif for the speech, and his use of the frontier metaphor, recalled the optimism of the Bush report, the Hoover speeches of the 1920s, and cultural attitudes toward progress running deep into the nation's past. On the importance of science and technology to the nation, the president and Congress seemed in accord. Both strongly backed programs to increase support for research. Frank Press, president of the National Academy of Sciences, noted in his annual report in April 1983, "My report to the members this year must be made in the context of a new mood in the country . . . one that looks to science and technology for economic progress and national security to an extent that may be unprecedented." And William D. Carey, executive officer of the American Association for the Advancement of Science, editorialized to his *Science* magazine readers, "With such blessings from both the executive and legislative heavens arriving in profusion after years of dwindling rations, scientists and educators alike could be pardoned for pinching themselves." [59]

President Reagan's famous "star wars" speech of March 23, 1983, in which he called for research to explore the possibility of an effective defensive system against ballistic missile attack, was of a piece with the State of the Union address. High technology would pave the way toward a more secure future for all Americans, and if the research effort were to be successful, the president would share the technology with other nations so as to make nuclear weapons potentially obsolete.[60]

Although his intent was to achieve control over the most threatening military technologies, Reagan's means for doing so involved reliance on high technologies and the nation's capacity to integrate and manage technological systems of unprecedented complexity. This is not the place to debate either the practical or the theoretical issues raised by the strategic defense initiative—whether its R&D programs could be administered by one central office, whether defensive systems could remain defensive or

Table 5-2.   **Defense and Nondefense R&D, by Type of Work, Fiscal Years 1980, 1988[a]**
Budget authority in billions of dollars

Type of work	1980 actual	1988 estimated	Percent change	
			Current dollars	Constant dollars
Defense R&D	15.0	40.3	169	83
Basic Research	0.6	0.9	64	11
Applied Research	1.9	2.6	38	−7
Development	12.5	36.7	194	99
Nondefense R&D	16.7	18.8	13	−24
Basic Research	4.2	8.6	107	40
Applied Research	5.0	6.5	29	−13
Development	7.5	3.7	−50	−66
Total	31.7	59.1	. . .	. . .

Sources: Albert H. Teich and others, *AAAS Report XIII, Research and Development, FY 1989* (Washington: American Association for the Advancement of Science, 1981), p. 8.
   a. Includes conduct of R&D only. Columns may not add because of rounding. Percentages based on unrounded numbers.

would become new destabilizing offensive systems, or whether the president had in some sense undermined the strategy of nuclear deterrence by staking out the moral high ground for defensive systems. What is noteworthy here is Reagan's embrace of technology as the solution to major problems of national security and economic progress.

This was perhaps the high point, midway through the first term, of the administration's faith in science and technology, a faith that did not erode completely but that became tempered by the realities of governance. It was almost as if the administration recapitulated the entire postwar experience with science and technology: optimistic faith followed by some disillusionment and then by a more realistic appreciation of what science could contribute to the solution of the nation's problems.

Table 5-2 summarizes the administration's changing emphases in R&D expenditures. The increased emphasis on defense is clear enough: defense R&D more than doubled, from $15 billion in fiscal 1980 to an estimated $40.3 billion in 1988. The sharp growth of basic research is also evident: nondefense basic research funding went from $4.2 to $8.5 billion. The defense contribution to the overall increase in basic research support, however, was a modest 11 percent in constant dollars as compared with a 40 percent increase for the civilian agencies. Development programs received the largest share of the increase in defense R&D, reflecting the growth in the acquisition of weapons systems. Development had always been the largest component in the defense R&D budget, but it had now increased from 83 percent to more than 91 percent of the total budget authority.[61] Basic research, 4 percent of the defense R&D budget in fiscal 1980, dropped to 2 percent by 1988.

The increase in the development component in the defense R&D budget reflected the administration's belief (more pronounced in the first than in the second term) that sharp distinctions based on definitions of appropriate public sector and private sector roles must guide decision-making. Since the Defense Department is the customer of technology development in its area, the government can appropriately fund as large a share of the development costs as it deems necessary. The effort to stimulate technology development in the civil sector generally, on the other hand, is not a government responsibility (at least not a DOD responsibility).

The civilian R&D agencies' function is to fund basic research and those few applied efforts that will not normally be adequately supported by market incentives. Hence the sharp drop in the development and, to a lesser extent, in the applied research category for civilian agencies in table 5-2.

This pattern illustrates why the Reagan administration, in its early policies, was not a mere replay of the postwar consensus. The consensus had declared an important role for the mission agencies—the Defense Department in particular—to support mission-oriented basic research. The Defense Department was also expected to take an expansive view of technology development in the boom years of the 1950s and 1960s. Attention was to be focused on innovations the military could use, yet the DOD would also broadly support the transfer of technology into the civilian marketplace. Most military contracts, for example, allowed the contractor to spend a small portion of the contract amount on independent research and development (IRD). This would enable the company to strengthen its technical capabilities, explore new products and systems before making a formal bid, and undertake related research activities that would contribute to the technical base underlying the nation's defense effort.

The Reagan administration initially opposed increased support of basic research by the Defense Department, following the premises of the Mansfield Amendment (which had disappeared from the statute books but not from the thinking of many policymakers). The administration also fought the practice of allowing independent research and development. Later it backed off from both practices. The DOD initiated a program to upgrade instrumentation in university laboratories. The military services helped modernize the manufacturing technologies and production systems of their defense contractors.[62] But it was not enough simply to change declared policy: altering the behavior of a bureaucracy requires active leadership.

A defense acquisition system that was once freewheeling gradually became enmeshed in controls and layers of bureaucratic review. Security regulations further clogged the system. In 1982 President Reagan signed Executive Order 12356, which reversed a thirty-year trend and tightened the rules on the declassification of research to guard against leakage of technological secrets. In this respect the administration aroused the fears of many scientists that Vannevar Bush's insistence on open communication of scientific ideas, which had seemed sacrosanct for so long, was now imperiled. Once in place, the restrictions acquired a momentum of their own; among other consequences the transfer of technology from the military sector to the civilian economy was slowed even when some logical application could be foreseen.

There were many paradoxes in science policy at the twilight of the Reagan presidency. The linkages between the affairs of science and of broader politics were especially complex. None has seemed more refractory and resistant to clear definition than the matter of innovation and its relationship to the competitiveness of American industry.

Policies toward Innovation

The Reagan administration's initial view was that the best policy toward innovation was to minimize the government's role in the economy. Thus it slashed the budget for Carter's energy demonstration projects, cut back the Commerce Department's office responsible for innovation policy, and otherwise sought to reorient budget priorities. The administration also sought "to stimulate industrial R and D and technological innovation by clarifying antitrust policies, reforming patent laws and procedures, streamlining regulations, and providing tax incentives."[63]

The tax incentives were embodied in section 221 of the Economic Recovery Tax Act of 1981, which provided for an R&D tax credit of 23 percent on new research expenditures, and in section 222, which provided a credit for gifts of scientific equipment to universities. Incentives were also provided in the accelerated depreciation allowance and investment tax credit provisions of the act.[64] In patents, the administration built on the reform enacted in 1980 and 1982 (the University and Small Business Patent Act and the Small Business Innovation Development Act and their subsequent amendments). Patent policies eventually were broadened into a component of the more comprehensive concept of "technology development," which included efforts to stimulate the transfer of technologies from the national laboratories to the marketplace.[65]

Antitrust policy was also an important element of the administration's innovation strategy. Policy was implemented by the simple expedient of bringing fewer cases to court, especially those involving collaboration among companies in R&D and in product development. The architects for this reformulation of antitrust policy were Stanford University law professor William Baxter, who served as assistant attorney general for antitrust from March 1981 to December 1983 and D. Bruce Merrifield, assistant secretary of commerce for productivity, technology, and innovation, who served through to the end of Reagan's term in office and into the Bush administration. In 1984 administration policy was placed on a statutory footing in the National Cooperative Research Act, which was intended to make it easier for firms to cooperate in research in the precompetitive stage. The act created a mechanism under which industrial research consortia could register their formation with the Department of Justice and thereby avoid treble damages if the venture were later held to violate antitrust law. Furthermore, joint R&D ventures under the act were to be judged on the basis of a rule of reason and not as per se violations of federal or state antitrust law.[66]

Regulatory reform was initiated by the creation of the Task Force on Regulatory Relief in January 1981, chaired by Vice President George Bush. Executive Order 12291 spelled out the broad principles that agencies were to use in determining whether their regulatory practices could be made less burdensome to the private sector.

Some observers have seen this cluster of policies as reflecting a tightly orchestrated effort by corporate elites and their allies to dominate decisionmaking and advance private interests. David Dickson, writing from the perspective of the European Left, found evidence that "decisions ranging from the broad allocation of scientific resources among competing areas of basic science, to the detailed application of scientific results to market-determined needs, are increasingly concentrated in the hands of a class of corporate, banking, and military leaders. . . . This process has been actively encouraged by the 'science policy' of recent U.S. administrations."[67] He believed, in short, that the tax, patent, antitrust, and regulatory policies of both the Carter and the Reagan administrations were designed to achieve corporate ends.

The conclusion reached by Harvey Averch seems closer to the truth:

At no time in the history discussed here has there been tough, critical, systematic third-party analysis of the proposed policies and the means of implementation. And at no time has there been consistent comparison of alternative strategies and their costs. The U.S. search for an innovation strategy has been marred by faulty design or, more accurately, by no design. Analysis has usually been placed in the hands of those with something to gain or lose. Alternative strategies have not been articulated or debated clearly, and values, facts, and predictions have never been clearly distinguished.[68]

Ideological consistency was not complete even at the beginning of the Reagan administration. While cutting the many Carter energy demonstration projects, for example, the administration fought hard to save the Clinch River Breeder Reactor project. The administration disliked, in particular, solar energy—a favorite of President Carter's—and strongly supported nuclear energy, considering the antinuclear bent of the Carter administration unreasonable and even bizarre. The Clinch River project was also strongly backed by Senate Majority Leader Howard Baker, who kept the project alive until it was finally defeated by the Senate on October 26, 1983. The 56–40 vote came on an amendment to the Supplemental Appropriations Act.

The evolution of energy R&D policies has been exceedingly complex, and confounds the application of any easy ideological formula. The Reagan administration backed fusion research (on the theory that its payoff was distant and therefore did not compete with commercial development) and initially also the Synthetic Fuels Corporation (as insurance against disruptions of oil supplies). The decisive event that shaped energy policy was the collapse of world oil prices, which began in 1981 and accelerated in 1982, removing most of the incentives for synthetic fuels and other energy demonstration projects.

The collision of ideology with reality was not limited to energy R&D policy. Consider how the administration modified its approach as it grappled with practical problems in NASA's aeronautics research program. Initially believing that aeronautics research was the responsibility of the aircraft industry, Science Adviser Keyworth convened a panel of the White House Science Council to consider a possible phaseout or modification of this program. The panel recommended that the program be maintained and even strengthened.[69] In fact, the linkages between the government and the aircraft industry dating back to the 1918 establishment of the National Advisory Committee on Aeronautics, the pooling of patents ordered then by Assistant Secretary of the Navy Franklin D. Roosevelt, and the subsidies of the 1930s for carrying the mail contributed greatly to the industry's development.[70] The review by the White House Council and OSTP staff confirmed that there was a range of technical questions concerning safety, aerodynamics, and common-use technologies receiving insufficient attention through the normal workings of the marketplace.

Yet no administration, beginning at least with President Kennedy's, has ever been fully satisfied with its mix of innovation policies. Officials have been unsure at which points along the stream of innovative activity public intervention would be most effective. They have also had difficulty in separating the effects of specific innovation policies from the context of macroeconomic policy as a whole. The Reagan administration, seeking to return to a pure market approach reminiscent of the Eisenhower era, found itself enmeshed in the web of initiatives launched by more recent administrations. It was pulled back from its doctrinal purity by precedent and by the program remnants from the interventionist strategies of the Carter administration.

Congressional pressure to act mounted as the administration confronted the recession of 1981–82 and then the growing competitive challenge from abroad. The need to do something (or at least to appear to be doing something) to assist American technological development beyond pursuing the correct macroeconomic policies and supporting basic research became a political imperative.

The administration's general strategy at its mid-point was perhaps best summed up in the January 1985 report of the President's Commission on Industrial Competitiveness, *Global Competition: The New Reality*. Under the chairmanship of Hewlett-Packard's chief executive officer, John A. Young, the commission pointed to such danger signals as the U.S. loss of world market share in high-technology sectors and called for a variety of cautious measures to strengthen the educational system at all levels, retrain the work force, and improve physical and human capital resources, as well as to pursue appropriate tax, trade, and antitrust policies.[71] The report clearly sided with the advocates of an open world trading system and resisted calls for protectionism. However, as if to illustrate the difficulty of charting an unambiguous policy course, the background volume of supporting studies tilted toward a strategy of greater intervention in trade, technology development, and sectoral policies.[72] As a result the report received a cool White House reception. The commission subsequently evolved into the privately funded Council on

Competitiveness, still chaired by Young and broadly representing business and labor viewpoints.[73]

The report contained one bold suggestion largely reflecting the viewpoint of presidential science adviser George Keyworth: it recommended the creation of a cabinet-level Department of Science and Technology. "Such a Department," the report said, "would make clear the importance of science and technology at a time when technological innovation is key to enhanced competitiveness. It would transform the current fragmented formulation of policies for science and technology into one that would be far more effective in meeting long-term national goals."[74] The proposal ran into the traditional opposition to a centralized science department. Such a move would, it was feared, create a large bureaucracy built up of program elements torn out of the context of the user agency and would therefore be likely to impede rather than hasten technological applications. Since science and technology funding was involved in so many different departments, any effort to centralize them would create endless jurisdictional disputes (equivalent, say, to centralizing all government lawyers or economists in one department). The proponents did not, in any event, make a convincing case that a conglomerate department would contribute to more effective mobilization of the total natural technology base. The idea died aborning.

In his second term, President Reagan moved steadily in the direction of pragmatic accommodation with those who wanted the government to do more to foster innovation, yet he did not abandon his basic philosophy. In the Commerce Department, for example, Assistant Secretary Merrifield pursued policies remarkably similar to those of his predecessor, Jordan Baruch, in the Carter administration. He acted as a kind of advocate of innovation within the government, making agencies aware of how their policies and practices affected industry. He also attempted to promote collaborative research among firms and disseminated information on recent process improvements in American manufacturing companies. His was an upbeat message that America could compete successfully in world markets.

The administration also supported passage of the Federal Technology Transfer Act of 1986 to expand the commercial transfer of R&D conducted in federal laboratories.[75] By executive order on April 10, 1987, the president implemented the act and gave special impetus to the idea of scientific exchanges between government laboratories and industrial laboratories. He directed the Departments of State and Commerce to devise a mechanism to make foreign technical information more widely available.[76] The act amended the 1980 Stevenson-Wydler Act, giving federal laboratories the authority to engage in cooperative R&D agreements with firms and consortia. The administration also supported passage in late 1986 of the Japanese Technical Literature Act, which gave the secretary of commerce the responsibility to acquire, translate, and disseminate technical information from Japan.[77] The potentially most far-reaching measures undertaken in Reagan's second term, however, were the technology-promoting provisions of the Omnibus Trade Act of 1988, adopted as a result of bipartisan support from the administration and the Democratically controlled One-hundredth Congress. Title V, sections 5112–5115 of the act changed the name of the National Bureau of Standards to the National Institute of Standards and Technology and gave the new institute broadened powers to assist industry to develop technology, to modernize the manufacturing base and improve product quality, and to commercialize new technology rapidly.[78] Section 5121 of Title V provided for technology extension activities to enhance productivity and technological performance, and section 5131 established an advanced technology program to assist in the commercialization of significant new scientific discoveries. Sections 5142 and 5143 created special review procedures and reporting requirements for monitoring developments in the semiconductor industry and in the field of superconductivity.[79]

The administration also stiffened its posture on trade and protecting intellectual property rights, espousing a doctrine of fair trade and attempting to head off more drastic action on trade policy by congressional Democrats. A task force appointed by the secretary of labor and chaired by former Under Secretary Macolm Lovell recommended a $1 billion effort to retrain displaced workers; many elements of their proposal were eventually incorporated into provisions of the 1988 trade bill.[80]

In 1987 the administration proposed Sematech, a six-year, $1.5 billion effort aimed at advancing the manufacturing technology underlying the U.S. semiconductor industry. The intent was to use federal funding, some $100 million annually from the Department of Defense, together with matching contributions from private industry and state and local governments, "to improve the equipment, materials, and techniques involved in the manufacturing process, as opposed to improving the design of semiconductor devices themselves."[81]

In the passage of the Tax Reform Act of 1986, support from the president and a bipartisan congressional coalition preserved the limited R&D

tax credit in the face of contrary pressures from tax reform enthusiasts (but the credit was reduced to 20 percent from the previous 25 percent and was extended only temporarily).

The Reagan administration continued to advocate a variety of technology initiatives—the space station, the superconducting supercollider, the new NSF science and technology centers—within the broad rationale of competitiveness. The proposals were so numerous and so redolent of a once-disdained interventionist strategy to aid the market in technology development that long-time Reagan critic and industrial policy advocate Robert Reich of Harvard's John F. Kennedy School of Government felt moved to chide the administration for excessive embrace of industrial policy goals.[82]

The Reagan administration, in short, struggled to balance the objectives of primary reliance on the market with selective intervention to provide incentives, to remove impediments, and sometimes to provide direct support for technology development. In this quest to reconcile partially conflicting objectives, ideology often gave way to pragmatism. This in turn led officials to struggle to maintain a consistent direction in policy. Answers to some of the most fundamental questions have remained elusive. What effect does the climate of government activity have on the pace and direction of innovation in the private sector? Does direct support from the government displace or induce additional private industrial research? Do industry and universities cooperate more effectively with, or without, government as a third partner?

The object of public policy—in this case the innovative behavior of companies—was itself imperfectly undestood. At a minimum the Reagan administration believed that the answers given by the Carter administration were ideological and had to be thought through afresh from a perspective that was not biased against the market. There was no clear pattern, with clearly understood steps, by which technology moved from the laboratory to a commercial product. Nor could one speak of any optimum relationship between the engineering, development, manufacturing, and marketing functions of the company—either in general or for specified industries. Nor was it clear how important technology was to the firm's product cycle, earnings potential, marketing strategy, and performance. The critical issues were the appropriability of technical advances for the firm (or the nation) doing the development work and whether a successful imitator could readily capture the benefits of an innovation.[83]

Figure 5-1. **R&D Expenditures, by Industry Group, 1964–85ª**

Constant 1982 dollars in billions

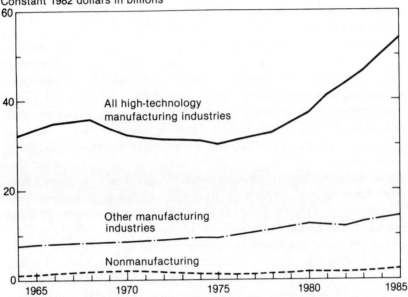

Source: National Science Board, *Science and Engineering Indicators, 1987* (Washington: National Science Foundation, 1987), p. 105.
a. Data include federally funded R&D centers administered by industry.

It was not easy to measure how well the patient—American industry—was actually doing. American companies as a whole in the early 1980s continued the strong growth in R&D expenditures that began after 1975, in part stimulated by the increases in defense spending (figure 5-1). But after 1984 the rate of growth in R&D expenditures slowed again in manufacturing (apart from high-technology industries, specifically primary and fabricated metals, petroleum refining, glass products, and wood products). In 1989 overall growth declined by nearly 1 percent in constant dollars, the first such drop since 1975.[84] Figure 5-2 shows the trends in industrial R&D spending.

Some parts of American industry made dramatic strides in improving manufacturing quality—the American automobile industry (and most particularly, Ford Motor Company) was a case in point.[85] But Japanese auto companies continued to display shorter cycles in the introduction of new models, lower engineering overhead, and a slight edge in product quality.[86] Productivity growth rates in U.S. manufacturing increased in the 1980s, but exclusive of manufacturing, gains have been substantially

Figure 5-2. **Expenditures for Industrial R&D, by Source of Funds, 1960–87**[a]

Constant 1982 dollars in billions

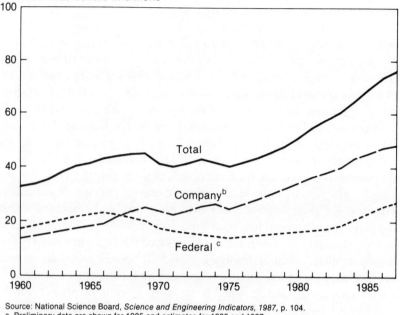

Source: National Science Board, *Science and Engineering Indicators, 1987*, p. 104.
a. Preliminary data are shown for 1985 and estimates for 1986 and 1987.
b. Includes all sources other than federal government.
c. Includes federally funded R&D centers administered by industry.

more modest, despite the increase of high-tech capital investment in services.[87]

Finally, one of the most difficult issues for policymakers seeking a coherent innovation strategy was to define the role of the Defense Department. The Reagan administration had started with the view that DOD was to play no part in nurturing the nation's civil technology base. In this assumption administration officials merely inherited the prevailing attitudes of the second phase of the nation's postwar science policy. Beginning with the decline of the defense share of overall national R&D expenditures after 1963, and symbolized in the Mansfield Amendment of 1970, the Defense Department was expected to focus on defense objectives. Part of the logic of the Mansfield Amendment was that the dominance of DOD in matters so far removed from national security considerations was unhealthy. A more rational allocation of roles was needed, with the civilian agencies responsible for nurturing the civil technology base. DOD's needs had become so specialized that they increasingly diverted technical resources away from potential commercial applications. Hence the drain on the nation's scientific and engineering personnel should be kept to the minimum necessary for national defense.

Accordingly, the Reagan administration initially opposed IRD and other technology-building strategies. But the obvious importance of increased defense spending to the nation's economy seemed to lead administration strategists inexorably back to assumptions more characteristic of the first phase of the nation's postwar science policy. DOD *should* play some role in strengthening scientific education, helping build the technology base, and promoting economic growth. The Defense Department therefore adopted the Manufacturing Technology program whereby the military services were to provide modest matching funds to encourage defense firms to upgrade their manufacturing capabilities. They also initiated a program to assist in modernizing university research instrumentation, eased their opposition to IRD programs, and embraced within the SDI program broad functions for aiding technology development.

Unfortunately, however, DOD's efforts frequently incorporated earlier policy conflicts without resolving them. SDI posed especially serious problems. Many of its objectives depended on fundamental research and basic engineering concepts—and hence were appropriate projects for the Defense Advanced Research Projects Agency (DARPA) or the basic research offices of the military services to support. Incongruously, the Defense Department classed such efforts under the 6.3, or advanced development, budget category and sought to focus university SDI research on specific development goals. In the heyday of DOD support for university research—when they might have been in a position to exercise such control—defense agencies did not tie research funding to specific development objectives. DOD has thus sought to return to early postwar patterns of nurturing university research while simultaneously pursuing policies reminiscent of the Mansfield Amendment era (that is, emphasizing more narrowly targeted research relationships with the universities).

Other policy conflicts have included the extent to which DOD should undertake programs such as Sematech that are intended to strengthen U.S. commercial interests. Observers have also queried whether defense procurement should be more open to foreign competition for contracts or to direct foreign investment in U.S. defense firms. The administration of U.S. export control, in particular the relationship between policies to promote exports and those to control technology flow, is a perennial dilemma now made more urgent by the revolutionary changes in the Soviet

Union and Eastern Europe. The layers of innovation policies since the Second World War have by no means left a Hegelian synthesis of the best previous practices. Rather, the legacy is one of contradictory impulses and imperfectly and only partially reconciled initiatives.[88]

Regulatory Reform

From the beginning the Reagan administration "sought to increase industrial productivity by relieving industry of unnecessary regulatory constraints."[89] This meant that efforts would continue along the three lines pursued by President Carter: implementing general economic deregulation, as in the transportation and energy sectors; improving the use of scientific knowledge in environmental, health, and safety regulations; and reducing unnecessary restrictions on the conduct of science itself. The goal of general deregulation was to abolish all rules that had no economic rationale—to allow, for example, free entry into air carrier and trucking routes and abolish price controls on natural gas. Science was to be used more carefully, not to abolish environmental regulations but to make them more efficient by identifying actual risks and weighing more accurately the costs and benefits of proposed regulations. Restrictions on science itself were to be reduced to ensure that it could continue to develop new techniques and tools of analysis.

Of course, there is considerable overlap among the three categories. The biotechnology industry, pharmaceuticals, and parts of the chemical industry, for example, are directly affected by constraints on scientific experimentation. Their growth is intimately bound up with the nature of regulations for toxic substances, carcinogens, and mutagens. Science itself is in a paradoxical position: it must be free from regulation so that it can improve the regulatory process. But if the conduct of science itself is dangerous to the environment or to the health of laboratory workers, it cannot be immune from regulation.

The underlying problem that faced Carter and then Reagan officials lay in the changing environmental threats and in the more sophisticated tools of scientific analysis for measuring the threats. In the era of the Delaney Amendment a few substances were assumed to be dangerous. More sophisticated measurements make it increasingly obvious that many environmental and health threats exist. While total bans were once seen as the solution to getting rid of an environmental hazard, the need for attention to the most urgent hazards in a world of varying thresholds of risk has become increasingly important. Setting priorities is complicated by the existence of mandated standards and legislative prohibitions, many of which have proved unattainable or attainable only at costs far beyond what Congress had envisaged. The challenge has become one of bringing standards more into line with what can be achieved but without sacrificing the interests of the public. In part, also, the problem lies in the successes of the environmental movement. The movement has achieved many of the simplest goals—cleaning up the obvious and easily detectable visible sources of pollution—and now faces more difficult-to-measure and chronic dangers that frequently show up only years after people are exposed to the hazardous substance.

Initial responsibility in the Reagan administration for the scientific aspects of regulation went to presidential science adviser George Keyworth, who chaired the Regulatory Work Group on Science and Technology set up under the Task Force on Regulatory Relief. The group included representatives of five regulatory agencies. Its charter was to improve the scientific basis for regulations, especially with respect to human carcinogens and toxic substances, and to upgrade the quality of government laboratory support for regulations and enforcement. At first many scientific regulatory issues were not of immediate importance to the president's top advisers. Key posts in the regulatory agencies initially went to inexperienced and ideologically inclined supporters. The Carter administration had exhibited a pattern of young, inexperienced regulatory enthusiasts held partly in check by scientists who advocated a more cautious approach and who urged that enforcement be linked closely to research. The early Reagan pattern was the reverse: that of a group of deregulators in the agencies directing the work of technicians who struggled to preserve the core regulatory functions. The combination of high-level inattention and ideological zeal at the operating levels was a formula for embarrassing failures.

The first step taken by the administration was in theory logical: bring legislative mandates into line with administration priorities through appropriate statutory changes. The Clean Air Act Amendments became a test case. The administration's proposals to relax a number of air quality standards were sharply rebuffed by Congress, under pressure of strong opposition by environmental interest groups. Even parts of industry, though anxious for regulatory reform, were unable to support fully the administration's moves. Extremists on both sides seemed to relish combat more than agreement.

The conflict continued on a second level with the administration's efforts to achieve reform through a greater OMB role in approving agency rulemaking. In particular, OMB imposed a requirement that environmental regulatory agencies adopt cost-benefit analysis as an operating procedure.[90] All environmental and workplace safety regulations were to be subject to a test that would apparently preclude action if there were minimal risk. A U.S. Supreme Court decision limited the applicability of the cost-benefit approach in OSHA-related matters, holding that congressional intent barred any weighing of general costs and benefits in regulations protecting worker health and safety.[91] The legal and policy disputes over cost-benefit analysis did not stop with the Supreme Court decision, however. The use of such analysis, and the activities of OMB's regulatory affairs unit, which monitored its implementation, exacerbated the conflict between Congress and the administration and triggered a long-running dispute over the limits of OMB power.

The most acrimonious environmental controversies in President Reagan's first term, however, centered on the actions of EPA Administrator Anne Gorsuch (later Burford). Initially the focal point of a dispute with Congress over the extent of executive privilege, she became embroiled in more and more controversies over environmental policy. In her twenty-two months in office she became the chief target of critics of the administration's environmental policies. She was forced to resign in March 1983 in the wake of a scandal involving Assistant Administrator Rita Lavelle (who was accused of political favoritism in altering the Superfund toxic waste cleanup list). With Burford's resignation the administration reached perhaps the low point of its political fortune during its first term.[92]

The crisis forced high-level attention to environmental regulation. In May 1983 William Ruckelshaus, who had served as EPA administrator from 1970 to 1973, returned. He succeeded in stabilizing the agency during his second term (which lasted until January 1985) and in bringing a more moderate tone to the administration's environmental policies. His goals were to devolve environmental decisionmaking to state and local levels where possible, to improve the scientific basis of regulations governing chronic health risks, and to narrow the differences between industry and the environmental movement.[93]

With Burford's departure, Interior Secretary James Watt, her longtime supporter and fellow Coloradan, became the focus of controversy. He remained the bête noire of environmental critics until his resignation in December 1983 following a furor touched off by a racially offensive remark at a Washington press conference.[94] Thereafter the key scientific regulatory posts were filled by a second generation of Reagan appointees in the Ruckelshaus mold. While they continued to clash with OMB officials and with the more militant White House domestic policy advisers, the appointees helped move policy toward accommodation with environmental and congressional critics. They were assisted by the fact that the administration, despite its strongly stated policies, did not give radical deregulation high priority.[95] Thus the more pragmatic officials were largely left alone to achieve whatever compromises they could with Congress until the level of conflict escalated to require presidential intervention.

The partial successes on environmental matters during the Reagan years included the reauthorization of environmental statutes after a period of executive-legislative struggle: the Resource Conservation and Recovery Act reauthorized in 1984, the Safe Drinking Water Act in 1986, the Superfund legislation in 1986, the Clean Water Act in 1987, and the Federal Insecticide, Fungicide, and Rodenticide Act in 1988. In addition, on January 7, 1983, the president signed the Nuclear Waste Policy Act of 1982, which set forth a plan for disposing of radioactive wastes. In several of the reauthorized statutes, Congress imposed tighter federal environmental controls. In the Resource Conservation and Recovery Act it mandated stricter standards for industry's handling of hazardous chemical wastes and broadened the coverage of the legislation to include tens of thousands of gasoline stations, dry cleaners, and other small businesses. In the Safe Drinking Water Act, Congress increased the number of contaminants to be regulated by EPA from some two dozen to eighty-three and set specific dates by which the agency had to issue standards. In the Superfund Amendments and Reauthorization Act Congress sharply limited the discretionary authority of executive branch officials and included state-by-state lists of specific dump sites that EPA was to clean up. In August 1988 the administration reversed a position taken in 1985 and indicated a readiness to limit nitrogen oxide emissions to levels proposed for 1987, thus clearing the way for the United States to participate in protocol negotiations with Canada, the Soviet Union, and European nations anticipated to lead to a treaty regulating acid rain.[96]

Despite environmental gains during the Reagan years, the policymakers have remained deadlocked on some critical issues. Industry and environmental groups, after seeming to have reached a partial modus operandi around the time of the Superfund reauthorization in 1986, have since hardened their positions. EPA budget cuts, especially for R&D, have made the basic goal of improving the scientific basis of environmental regulations more difficult to achieve. Short-run, enforcement-oriented

R&D efforts fared significantly better than the longer-term research.[97] It may have been understandable for the EPA to focus on enforcement, but a commitment to improved scientific understanding in the regulatory process cannot be sustained without adequate funding.

In general, the environmental regulatory agencies have had less capacity to defend their budget requests than larger agencies and even less ability to obtain expanded funding. The administration, though facing constant pressures to control expenditures, at times also appeared to believe that curbing the growth of regulatory staff and staff resources was a useful component of regulatory relief strategy. Undoubtedly, the administration also had priorities it regarded as more critical and simply did not want to take on any unnecessary battles. Mindful of the criticisms levied against President Carter for allegedly squandering political capital by taking positions on too many issues, Reagan's legislative liaison team sought to limit presidential intervention to a few urgent matters.[98]

The resulting pattern of policymaking meant that agency officials were left free to make policy if they could achieve accommodation among competing interests. Unfortunately, they often could not, and paralysis resulted. Environmental decisionmaking was so fragmented in Congress that only strong executive leadership could reconcile the intense conflicts. Without it, when OMB and the executive line agencies were opposed, and when congressional factions fought for jurisdiction, no one could agree on legislative mandates or the details of program implementation. Even when Congress acted to reauthorize an environmental statute passed during the heyday of the movement, the issuance of regulations was often held up in court action. The OMB, though losing a critical legal challenge in federal district court, continued to exercise broad authority to oversee agency rulemaking and hence to slow the process. When rules were issued, outside parties often instituted legal challenges. From one side, the typical regulatory agency was assailed for proposing an unenforceable rule to implement an ambiguous, unworkable, or unconstitutional legislative mandate. From the other side it was accused of failing to protect the public health, failing to comply with legislative intent, moving too slowly, and weakening the enforcement standards demanded by Congress. Whichever way the agency moved, the matter seemed bound to end up in court. The result was delay, an administrative process bogged down in judicial scrutiny and legal maneuvering, and often no clear resolution of the issue.[99]

The nation achieved some successes in protecting the environment without creating a regulatory regime that would suffocate economic activity. In this sense, policy reached some balance of forces. But continued stalemate in many matters offered little satisfaction to activist interest groups, to industries affected by regulation, or to the public. In the initial postwar period too little heed was paid to the control of technology. During the environmental movement, there was the belief that technological quick fixes could be found to any complex problem. Hence many of the environmental laws contained unattainable objectives that were ostensibly to be achieved by forcing the offenders to develop the best available control technologies. Once mandates were enshrined in law, retracing steps and redefining objectives proved difficult. The nation continues to confront the challenge of resolving these policy contradictions.

The Foreign Policy Context

When President Reagan assumed office, his main goal in foreign and defense policy was to build up the nation's military strength. He also wanted to halt what he saw as a dangerous erosion of U.S. influence in the world, as exemplified by Iran's holding American diplomats hostage. This orientation stemmed logically from his domestic priorities: cut the civilian side of government to stimulate economic growth through reduced taxes and thus allow a defense buildup.

While the administration's early preoccupations lay mainly in domestic politics and policies, there was certainly a distinct view of the world. Supporting its belief in the benefits of a military buildup were assumptions about the dangerous instabilities of the modern world, the assurance that peace came through strength, and a deep-rooted anticommunism.[100] The détente of the Nixon-Kissinger-Ford era and the Carter administration's embrace of arms control in the proposed Salt II Treaty were flawed because America had been lulled into forsaking aggressive pursuit of technological advantage. While slowing the growth of America's own technology, détente had allowed U.S. enemies to develop more rapidly through slack administration of the export control laws (in particular, through the leakage of sensitive technologies by our allies). Moreover, the Reagan administration believed, the arms control efforts had paid insufficient attention to verification problems. Consequently, Soviet violations had gone undetected or unreported or both. Furthermore, it was impossible to stop technological developments as completely as arms control advocates supposed. The nation should thus rely on the uninhibited growth of its military technology as the safer route to security. The allies were now technologically more sophisticated than they had been just after the war but they were indifferent partners in the administration of export controls.[101]

Reagan's outlook did not accept what seemed to be an almost theological belief in arms control doctrine on the part of some Democrats. The administration did not quarrel with the first significant step to control military technology, the 1963 Partial Test Ban Treaty. The Nuclear Nonproliferation Treaty of 1969 was also a logical limitation on the spread of nuclear technology that could benefit the nation. The doubts of Reagan strategists tended to focus on the ABM Treaty, which was the proudest achievement of the arms control proponents and the cornerstone of the whole effort to bring the nuclear arms race under control.

The ABM Treaty of 1972 had been undertaken on the logic that it was necessary to limit defensive systems before fixing an upper limit on offensive weapons and then gradually reducing the numbers of missiles, the intent of SALT I and SALT II. But critics from both the Left and Right pointed out flaws in the logic. The limits, if set high, could lead to an arms buildup. The development of new missile systems could lead to force expansions, and those forces would tenaciously cling to life. And disputes ensured ample challenge to the Soviet record on arms control compliance. The radar station at Krasnoyarsk was a favorite target of the critics in the Republican party and seemed to head any list of Soviet violations (indeed, in late 1989 the Soviet Union confessed to the violation).[102]

There was sufficient doubt about arms control matters in the minds of Reagan officials to justify a pause to reappraise doctrine while in the meantime strengthening the nation's defense. This increased strength would only come through unleashing the potential for advances in military technology. Defense and foreign policy would adopt budget, innovation, and regulatory policies to realize that potential. This was the path to be followed even if it meant a more assertive, unilateral role for the United States in the world.

There are inevitable pressures that draw presidential decisionmaking on defense and foreign policy toward the political center. The Reagan administration followed a moderate course on a number of important issues. For example, it maintained the opening to China by refusing to sell an advanced fighter plane to Taiwan despite strong sympathy for the idea in the right wing of the Republican party. The administration also announced that it would abide by the limits of Salt II even though it considered the treaty fatally flawed (the limits were exceeded finally in 1986 as U.S. Trident submarines began to be launched, but this was in a radically different context of arms control negotiations). Nor was there any instance of rash or reckless use of American military forces.

Nevertheless, the early Reagan policies followed a course clearly different from Carter and Ford policies, one that coincided with the conservative critique of the liberal establishment. The administration entered into a protracted quarrel with NATO allies over the sale of Soviet gas to Western Europe, which involved sending the Soviet Union details of pipeline technology and compressors that the administration regarded as having potential military applications. (In the process the administration asserted the extraterritorial reach of American law to American subsidiary firms operating in Europe.) The administration took America and a group of key allies out of negotiations on the Law of the Sea Treaty. It opposed international economic coordination of the sort that would attempt to manage exchange rates (consistent with its views of minimizing government intervention into the domestic economy). It gave notice of U.S. withdrawal from the United Nations Educational Scientific, and Cultural Organization (UNESCO) because of the organization's bad management and an alleged ideological tilt to the left in its activities. Finally, the administration announced a defense buildup projected to cost $2 trillion over five years.

The president's "star wars" speech of March 23, 1983, perhaps marks both the high point of this first phase of the Reagan defense policies and the beginning of a transition to more moderate policies. Coming just two months after the State of the Union address in which he spoke of the "technological frontier," the March 23 speech announced a research program intended to test the feasibility of a defensive system against ballistic missile attack. If successful, the program could lead to the development and deployment of a shield that would render nuclear weapons "impotent and obsolete." [103] The speech was later followed by a reinterpretation of the ABM Treaty that would permit the testing in space of defensive systems based on futurist technology. In depicting the strategic defense initiative as "the key to a world without nuclear weapons," Reagan was advancing his hope for and faith in a technological solution to a complex policy problem. In effect, technological advances would reverse the advantage that had seemed to lie with offensive systems since the advent of nuclear weapons.

Almost at once critics assailed the proposal from all sides. The president's idea seemed a caricature and an oddly discordant synthesis of the nation's postwar attitudes toward technology: the exaggerated faith that technology could by itself solve critical problems and the implicit assumption that nuclear weapons were an unparalleled evil and should simply be removed, like pollution, by forcing the polluters to apply the best available technology.

Reasonable opposing arguments were that it was not easy to move from laboratory concept to an operational system, that the costs of the new technology as well as its theoretical benefits had to be carefully weighed, and that the transition to a new form of social organization could upset established controls and limitations, with unintended consequences. To stake all on the possibility of breakthroughs in defensive ABM technology while unraveling the existing ABM Treaty might invite a new technological arms race. To define nuclear weapons as no longer critical to Western defense strategy might weaken the NATO alliance and portend new threats to peace and stability. The arguments seemed to replicate the entire postwar debate on the promotion, control, and coordination of science and technology as instruments of national policy. Nevertheless, the debate over SDI showed President Reagan's political genius. By advancing an apparently outlandish proposal, he disarmed the antinuclear critics, frightened the Soviet Union into a resumption of arms control talks, and utterly dominated the agenda of policy debate on strategic arms. Opponents, caught up in denouncing SDI, never advanced serious proposals of their own.

The debate over SDI began to produce a new political dynamic at home and to alter the international context. The administration modified its strategic arms negotiating position in a bargain with a group of centrist senators and representatives holding the balance of power in Congress; they in turn agreed to support the administration's advocacy of the MX missile.[104] The potential uses of SDI as a bargaining chip in negotiations also contributed to heightened interest in arms control within the executive branch. Whether any of the president's critics at home considered his SDI proposals feasible, Soviet leaders, perhaps fearing that the program would give an across-the-board boost to American military technology, began to revise their own approach to strategic arms negotiation. With the accession of Mikhail Gorbachev as general secretary of the Communist party in March 1985, the Soviet Union entered into an activist new phase of arms control negotiations with the United States.

It is not necessary to retrace the events leading to the December 1987 agreement between the United States and the Soviet Union on intermediate and short-range nuclear weapons and to significant progress toward an even more dramatic accord on deep cuts in strategic weapons. Reagan, criticized during his first term for being the only postwar U.S.

president failing to meet face-to-face with Soviet leaders, had held by the end of his two terms an unprecedented four meetings with Gorbachev, including a visit to the country he had once described as the "focus of evil in the modern world." [105] With the apparent U.S. endorsement of the Soviet leader and *perestroika*, superpower relations had reached a point that had seemed inconceivable a few years earlier.

The administration gave up the notion that security could be achieved only, or even mainly, through technology and returned to a belief in controlling weapons development through diplomacy. The president, especially after the Democrats regained control of the Senate in the November 1986 mid-term elections, was forced to pledge that he would abide by the narrow interpretation of the ABM Treaty. The administration had also moved closer to traditional multilateralism in other matters. Beginning in 1985, for example, with the effort of the Group of Five finance ministers to adjust exchange rates and trade imbalances, the administration looked more favorably on economic coordination. The attempt to sell the idea of SDI to allied governments also forced the administration toward a more conciliatory posture in alliance relations and elicited implicit pledges that controls on technology transfer, such as those it tried to impose during the gas pipeline controversy, would not be attempted again.

The generally moderate tone of the later Reagan administration in defense and foreign policy matters should not, however, lead to the conclusion that overall consistency in policies and clarity of political purpose were achieved. The movement away from doctrinal purity and toward the center, in fact, may have meant that the Reagan policies increasingly reflected the ambiguities, inconsistencies, and incongruities of mainstream American politics. The consensus on science policy had not so much been rebuilt into a coherent new whole as patched together, and it incorporated even more discordant elements than before. The accretion of unresolved problems from the past and the emerging challenges left an unfinished agenda of science policy for the nation.

The 1980s in Perspective

Postwar science policy was born in the effort to understand the implications of revolutions in nuclear and space technology for the nation's security. Almost half a century later, through the vicissitudes of cold war and détente and global economic competition, the issues posed by the intersection of national security and other national interests remained the central concerns of science policy. Defense and space-related R&D had increased its share of federal research support—not to the dominant levels of the early postwar years, or even of the 1960s, but significantly nonetheless. The role and impact of defense and space-related research on the nation's total research effort, on its scientific institutions, and on the personnel sustaining the national scientific capability were subjects requiring the most careful thought. The relaxation of East-West tensions and the prospects for reduced military budgets posed critical new intellectual and policy challenges.

While the Reagan era was a period of unusual growth in government R&D budgets, the results of that investment did not always match expectations. The nation's research system, while still healthy—and, indeed, on the threshold of remarkable advances in many fields—was, by the latter part of the Reagan administration, suffering from an accumulation of problems that prevented any new golden age of government-science relations. Signs of an erosion in the nation's scientific and technological capacities caused concern for the future, even as momentum carried the system along.

The future will bring complex choices on the structure of the research system and on ways of matching the system to the nation's needs. This was clearly illustrated in the debates over R&D budget priorities in the final year of the administration. In November 1987 the budget "summit" conference resulting from the stock market collapse of October 1987 resolved the deficit issue for the fiscal 1988 and 1989 budgets by devising ceilings for expenditures in the categories of defense, international, domestic, and nondiscretionary domestic programs.[106] Domestic expenditures for 1989 could only increase to $147 billion (from the $144 billion in 1988). Notwithstanding the constraints, the administration proposed increases totaling $2.7 billion for R&D alone in the fiscal 1989 budget. Large civilian projects accounted for the lion's share: the space station in NASA's budget was projected to increase its obligations to $739 million in 1989; the superconducting supercollider in the Department of Energy to increase to $430 million; a new NSF initiative for science and technology centers at universities was assigned an initial outlay of $130 million.[107] Other projected increases were for NASA, for the plan to try to double NSF's budget in five years (the fiscal 1988 request for a large NSF increase was substantially scaled back by Congress), and the Department of Energy's general science programs.

Thoughtful observers doubted that R&D requests of this magnitude would be approved by Congress. Competition between R&D and other expenditures would be intense, and not all of the proposed new starts in scientific and technological programs were likely to be funded adequately in the out years. R&D programs generally seemed in danger of becoming Pentagonized, of suffering the fate of too many system start-ups, each with a tenacious constituency fighting for resources. The result could be a great many partially funded and crippled systems. The matter of priorities, it seemed, could no longer be avoided.

Frank Press warned members of the National Academy of Sciences in April 1988 that they could no longer take refuge in the belief that they "should support all of the science initiatives on the table," because to propose a list of priorities would only serve "to divide [the] community and to insure a reduced budget." Priorities were needed to maintain scientific progress and to quell unseemly internecine conflicts over funding. Even with a political consensus for generous support of research, funding would be insufficient in the Gramm-Rudman-Hollings era for all the grand plans of American science. The case would also have to be made for research programs in the total context of public policy and of larger social needs. "It is feckless and destructive," Press said, "for the scientific community to argue 'for science at the expense of the homeless,' as one member of Congress put it. It is also unrealistic to argue for both science and social programs until the deficit is brought under control." [108]

Two weeks earlier, Robert M. Rosenzweig, president of the Association of American Universities, told an audience of science administrators and university officials at the Thirteenth Annual AAAS Colloquium on R&D Policy that he could "see no realistic set of circumstances in which science will receive from the federal government all that it wants, needs, or deserves." If it was both "necessary and proper to make hard choices in the defense establishment," he asked, "does that tell us anything about the necessity and possibility of doing the same in the science establishment?" Rosenzweig asserted, furthermore, that his association had "looked hard at the need for assistance in reconstructing the capital base of university research and concluded that the need is so great that it is worth doing even if funds for this purpose were to reduce project funding." [109]

Thus in President Reagan's final year in office the administration and the scientific community found themselves in a strange reversal of roles. Scientists began to call for more setting of priorities and criticized the administration for embracing scientific efforts too indiscriminately. The "big science" program came in for special criticism. When budgets were projected, the large, expensive activities threatened to squeeze out sup-

port for the smaller activities, jeopardizing the underlying human resource base of science. The administration, having backed off from its
earlier plans for a more selective pattern of research support, now embraced scientific advance as the answer to the lagging productivity
growth rate and to the nation's problems of insufficient economic competitiveness. It invoked the unexpected synergies of research, along with
the need to have patience and to avoid excessive "targeting" so that long-
term benefits and spillover effects from research could be realized. These
familiar arguments had been, in fact, the stock in trade of the scientific
community for years. The administration and the OMB thus stood fast
as defenders of science's autonomy while the scientific community urged
the executive branch to be more realistic in assessing the ability of research to contribute to the solution of national problems.

The new realism within the scientific community reflected two trends.
First was a growing conviction that science is genuinely at a kind of
crossroads where hard choices have to be made to ensure its future
health. Support for science had become such a complex issue, many
scientists felt, that the nation could no longer afford the comfortable
evasions and equivocations that had often passed for thought from policymakers and scientists. They were also less inclined to make promises.
Second, direct appeals to Congress, which seemed a natural recourse
when the Nixon administration decapitated the science advisory system
within the executive branch, had lost much of their lustre in light of the
growth of pork barrel politics in congressional decisions about scientific
facilities. These appeals could jeopardize one of the most critical elements of the whole postwar system—the presumption that objective criteria and the judgments of scientific peers would be prominent in decisions about research support. Lawmakers and scientists from "have not"
institutions dismissed these fears as the anachronistic longing of members of elite scientific institutions who wished to preserve their privileged
position in a closed decisionmaking system. It was, in fact, clear that
Congress had acquired power in science policymaking that it would not
surrender and that the demands for more geographically equitable funding of science would intensify.

As the 1990s approached, the nation stood at a turning point in the
postwar history of efforts to relate scientific and technical resources to
human and social purposes. Reagan seemed first to have embraced science and technology with zeal and with a clear vision, then to have made
pragmatic accommodations as doctrine collided with reality. In the end
the Reagan era left an unfinished agenda of science policy concerns and

the task of defining a new vision to guide the relations between science and government.

It was the best of times for science and the worst. Scientific disciplines enjoyed spectacular successes in advancing the frontiers of knowledge. Support for basic inquiry was at unprecedented levels. American industry apparently was strengthening its technical base. But developments within science and within society precluded a return to the simple faith of the Bush report. Critical choices confronted program planners, science administrators, and elected and appointed officials. Set on its present course, the nation could drift into the position of excessive faith in science and technology, of undertaking too many initiatives that could not be successfully completed, of working at cross-purposes with itself. Scientists welcomed the chance to serve society but felt increasingly uneasy over the terms of the bargain between science and society. Science had moved, as Vannevar Bush had wished, from the wings closer to center stage. But with scientific affairs and broader public affairs so deeply intertwined, a possible outcome was that science policy would lose its distinctive focus. Could a new framework for government-science relations be devised? Or, if that were too grand a goal, could the nation at least resolve some of the present policy conflicts? How could America achieve a more satisfactory synthesis of the differing principles at work since 1945?

6 ||| The Unfinished Agenda

THIS BOOK HAS TRACED America's science policy through three eras. Just after World War II science and technology, having ushered in spectacular changes in warfare, transportation, communications, work habits, and health care, were hailed as holding the key to the future. In the 1960s and 1970s optimistic hopes gave way to a more realistic assessment of science's contribution. At the same time, the nation sought to channel research and development toward a variety of social ends—a cleaner environment, safer transportation, more civilian jobs, better health for all Americans—and weapons, space exploration, and atomic energy became less important matters for science policy to address. Under President Reagan, the nation moved back toward some of the earlier attitudes: emphasis on national security, faith in basic science, less fear of adverse side effects from the growth of technology.

Science policy has been discussed at several levels in the preceding pages. The merits and consequences of the substance of particular policies have been examined. The policy process has also been discussed. How did decisions get made? What were the effects of the ways various groups pursued their goals? The policies directed toward science-related problems and the process by which issues are addressed add up to a system of institutional roles. Assumptions about the responsibilities and behavior of various agencies and actors are part of the system. Hierarchies of values and goals guide the relations between scientists and others. The overall performance of the system has been an underlying concern of this inquiry. Has the nation been well served or poorly served by the system devised to support and to make use of science?

Maintaining Pluralism

The nation has generally been well served by the system of determining science policy. It has maintained a measure of continuity as it has

Table 6-1.   **Share of Total R&D Funding, by Sector, 1967, 1971, 1976, 1984**

	1967		1971		1976		1984	
Sector	Billions of dollars	%	Billions of dollars	%	Billions of dollars	%	Billions of dollars	%
Federal intramural	3.4	15	4.2	16	5.8	15	11.6	12
Industry	16.4	71	18.3	69	27.0	69	71.5	73
Universities	1.9	8	2.5	9	3.7	9	8.5	9
Federally funded R&D centers	0.6	3	0.7	3	1.1	3	3.1	3
Other nonprofit	0.8	3	0.9	3	1.4	4	3.0	3
Total	23.1	100	26.6	100	39.0	100	97.7	100

Source: National Science Board, *Science and Engineering Indicators,* 1987 (Washington: National Science Foundation, 1987), table 4.4.

adapted to new demands. Universities, government laboratories, and industry continue to perform functions similar to those they performed half a century ago, despite vast changes in national priorities. Their relative shares of federal and of total national R&D funds remain fairly stable (table 6-1). The only significant departure is the increase in industry's share of total R&D funding that occurred in the Reagan era.

Furthermore, when national priorities have changed, universities, government laboratories, and industry have adapted with the times. But they have done so gradually, which means that change has been absorbed without disruptions or major breakdowns. Challenges to institutional functions and the overlap of functions have always been present; the system, it seems, has always had some loose ends and running disputes over the values to be served.

In his organizational recommendations, Vannevar Bush envisioned something very close to a Department of Science that would be responsible for virtually all research and generic technology development, leaving only specific development in the hands of the mission agencies. He sought this centralization because he believed it would best promote the progress of science (an instrument serving social ends and a good in its own right). Senator Harley Kilgore also favored a Department of Science but for other reasons. He believed that the diffusion of knowledge and its practical applications would be fostered if a single accountable official were in charge of allocating funds for research and development. The nation, in effect, decided finally by means of political compromise, without ever explicitly debating the matter, that generating and diffusing knowledge were both desirable and that one controlling vision was to be

Figure 6-1. **R&D Expenditures, by Character of Work, 1960–85**

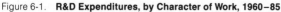

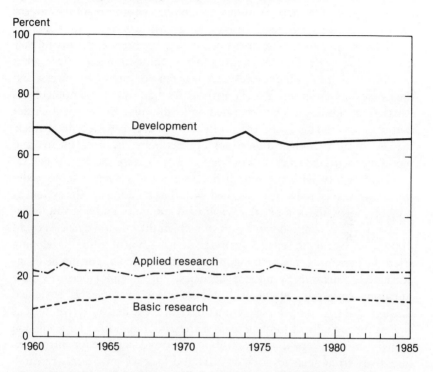

Source: National Science Board, *Science and Engineering Indicators, 1987,* table 4.6

eschewed in favor of a plurality of approaches. A biomedical research empire would serve the needs of public health and medical technology, defense agencies would foster research and development related to defense needs, and energy research would be supported as an instrument to advance the goal of low-cost clean energy. A smaller central agency would attend to scientific personnel and to those fields not adequately supported by the mission agencies. The whole system would tilt toward development rather than basic research, but research would be adequately supported as an overhead cost on development. Figure 6-1 shows the stability in the shares awarded development, applied research, and basic research in national R&D funding since 1960.

Despite recently revived interest in a Department of Science and Industry, the arguments for a major restructuring of the federal system of research support do not seem compelling.[1] The principal arguments are that a science department would mean more bureaucratic clout for science activities (and thus higher priority for them in budget battles), that it would improve America's ability to mobilize efforts to address urgent priorities, create synergies across existing jurisdictional lines, eliminate duplication of effort, and create a civilian counterweight to military domination of R&D expenditures. The counterargument states that scientific activities in themselves have no claim on public resources, and the nation has supported R&D only because it advances other objectives. R&D thus should not be separated from the missions of the agencies that support it. This is the reason there has never been a federal R&D budget as such; budget allocations are made by the departments and represent their assessments of the best means to accomplish their goals.

The difficulty with the pluralistic system arises when one concedes that some agency must be concerned with the soundness of the system as a whole, for without a healthy scientific community and a strong infrastructure, there would be no base from which the mission agencies could draw. The National Science Foundation is responsible for developing scientific personnel, maintaining infrastructure, and supporting fields inadequately nurtured by mission agencies, but it cannot perform these functions by itself. Biomedical research, for example, is sustained through NIH funding for personnel development and infrastructure as well as through direct funding for research. NIH and other large mission agencies must accept this responsibility. To take support for infrastructure away from agencies such as NIH and to leave only direct research funding would be inefficient and would create more administrative confusion than exists now.

To combine NIH with other agencies into a new department would gain little if it meant merely a paper reassignment of responsibilities and official titles.[2] But if a real change of missions and responsibilities were involved, a protracted and bruising battle among the constituencies affected could be anticipated. The prospects for success would be uncertain, and there could be a long period of administrative confusion and near chaos. The reasons the nation did not create one large science department are still relevant. Those who would be affected are unwilling to trust a single overriding vision on the proper goals of science policy, preferring the deeper rationality of multiple and partial visions of the common good. There is also fear that a large department would embrace too exclusively either Vannevar Bush's goal of generating knowledge or Harley Kilgore's goal of diffusing it. If the department had no clear goal or if its goals were seriously in conflict with one another, the difficulty of achieving balance and sensible trade-offs would be compounded.

Further, aggregating many research programs under one organizational structure could result in budget cuts rather than increases, particularly in the context of the Gramm-Rudman-Hollings budget restrictions, which seem likely to persist. Undoubtedly, a need exists to examine national interests that cut across departmental lines and evaluate how well different agency programs fit together and how well they serve those national interests. But this should be done by the White House with the aid of the president's science adviser, not at the departmental level where the contending interests are fighting for their own causes. The executive agencies cannot be both in the battle and above it. No one department can act as the president's agent in resolving policy disputes. Nor can it oversee program implementation and coordination. Interagency coordination must be done by the White House in close integration with the budget process.

Even with strong White House support, individual programs can be neglected in a large department. President Reagan's strategic defense initiative, despite being one of the administration's highest priorities, received only modest incremental funding after becoming a separate program in the Defense Department. The fate of many science programs, if thrown together into one department, would depend on the priority given each by the departmental leadership. Once the constituencies supporting the programs became fully aware of the uncertainties of reorganization, they could be expected to wage a protracted fight against it. Thus while the advantages of reorganization to consolidate science programs are unclear, administrative turmoil, lengthy political battles, and confusion would be certain.

An issue of great importance raised by the advocates of a science department is the status of civilian R&D and the need for more balance between it and military research expenditures. The United States lags behind Japan and West Germany and is not far ahead of France and the United Kingdom in civilian R&D expenditures as a percentage of GNP (for 1985, 1.86 for the United States as against 1.85 for France, 2.53 for West Germany, 2.75 for Japan, and 1.71 for the United Kingdom—see chapter 4). But U.S. R&D devoted to defense relative to total R&D expenditures has consistently been the highest among OECD nations. Even before the Reagan defense buildup, America devoted more than half its R&D expenditures to defense, as compared with 49 percent for the United Kingdom, 39 percent for France, and 15 percent or less for Sweden, Switzerland, Germany, and Japan.[3] Earlier the percentage was higher (93 percent in 1963 if one counts space research as related to national security, a reasonable assumption in the period before the Apollo program). At the end of Reagan's presidency defense accounted for nearly 70 percent of federal R&D expenditures.

The proportion of public resources going to defense R&D is partially offset, however, by the magnitude and scale of civilian research expenditures.[4] Too much is sometimes inferred from the heavy U.S. investment in defense R&D, as exhibited in the thesis that the burdens of defense and world leadership lead directly to economic decline.[5] Both as a percentage of GNP and of total federal outlays, defense expenditures in 1987 were lower than at any time during the heyday of U.S. economic dominance.[6] It is not the burden of world leadership alone therefore that has created America's problems with competitiveness.

Clearly, however, the changing international economic environment and the altered needs of national security invite the attention of policymakers to trends in the allocation of R&D resources between civilian and military requirements. Policymakers want to know, of course, what level of defense R&D spending is needed, but also to what extent these expenditures are likely to stimulate the commercial exploitation of new technologies.

Important light is cast on the issues by Henry Ergas's comparative analysis of mission-oriented and diffusion-oriented strategies in national technology policies.[7] Three of the industrialized countries he examines (the United States, the United Kingdom, and France) have mission-oriented policies. They have a high percentage of defense R&D expenditures and emphasize a few high-visibility goals of national importance that represent radical innovations. The technologies they pursue are complex systems to serve the needs of a particular government agency, and the companies that perform the work are usually large and have sophisticated technological capabilities. There is some diffusion of technologies to smaller subcontractors, but it is as a by-product of achieving the primary objective and not as a goal of policy. In contrast, policy in West Germany and Sweden is primarily diffusion-oriented, emphasizing commercial exploitation of technology. But each national system exhibits instances of both diffusion and mission orientation so that the differences are not as sharp as they may at first appear. Nonetheless the diffusion-oriented countries seek to diffuse technological capabilities throughout the industrial structure, thus facilitating incremental adaptation to change. Japan is a unique mix of diffusion and mission-oriented policies (table 6-2).

The mission-oriented countries, most notably the United Kingdom in

Table 6-2. **Percentage of Total Public R&D Funding, by Country and Type of Industry, Estimated 1980**[a]

Country	High-intensity industry	Medium-intensity industry	Low-intensity industry
United States	88	8	4
France	91	7	2
United Kingdom	95	3	2
West Germany	67	23	10
Sweden	71	20	9
Japan	21	12	67

Source: Organization for Economic Cooperation and Development, in Henry Ergas, "Does Technology Policy Matter?" in Bruce R. Guile and Harvey Brooks, eds., *Technology and Global Industry: Companies and Nations in the World Economy* (Washington: National Academy Press, 1987), p. 194.

a. High-, medium-, and low-intensity R&D industries are defined as those whose ratios of R&D expenditures to sales are, respectively, more than twice, between twice and half, and less than half the manufacturing average.

Ergas's analysis, concentrate resources in too few hands; consequently there is less stimulus to industrial innovation. Certain features of the U.S. economy and its technological infrastructure mitigate the impact of its emphasis on mission. These include the sheer size of the U.S. research effort, which permits alternative design concepts and experimentation, at least in early stages of investigation, and the large pool of technological expertise (especially in the universities) that can be drawn on by government agencies without crowding out the availability of expertise for other applications. The mobility of the U.S. scientific work force and the relative openness of the administrative system, which tend to promote diffusion even in the absence of any conscious effort to do so, also have a mitigating effect. Only a quarter of federal R&D in America is performed by the government, so that public funds are inevitably placed in the hands of private contractors who in turn stimulate technological awareness in their suppliers.[8]

The United States does not entirely escape, however, the problems of mission orientation and heavy emphasis on defense. Large defense programs that have not proved successful are often allowed to run too long, while some civilian programs are abandoned prematurely. The pool of technical personnel, although large, is still limited. While the point is difficult to quantify, the potential diversion of the most talented scientists and engineers from civilian manufacturing may help account for the lagging quality and poor reliability of American products. Because 17 percent of U.S. scientists and engineers are now foreign-born, the nation is also potentially vulnerable to a brain drain.[9] Moreover, certain cultural features in the United States exacerbate the drawbacks of mission orientation, especially a short attention span that aims at dramatic break-throughs rather than incremental additions to knowledge or gradual improvements to products. This may explain why Americans are often first to introduce technology, but fail to capture the mass markets in the more mature phases of product development.

Ergas is cautious about drawing wide-ranging policy implications from his analysis, but several conclusions seem clear. The diffusion of technological capabilities across the entire economy, especially to technologically less sophisticated firms, deserves high priority even if the nation must continue to spend large sums on defense research. A large new mission agency charged with dramatic program goals in civil sector technology development provides no easy answer. Indeed, there is a high likelihood that it would tend to concentrate resources on a few politically chosen objectives instead of market cues. Building a wall between defense and civilian technology applications, as the Mansfield Amendment did, is the wrong solution. Drawing the DOD more fully into developing and diffusing civilian technology, as was envisaged in the Packard Commission recommendation for greater reliance on off-the-shelf technologies, is the preferable path.[10]

An Affordable Research System?

A growing sense of unease about the way resources are allocated has troubled observers of U.S. science affairs. For the first time since World War II, scientists have called for priorities to guide the allocation of R&D funds. Policymakers understand that despite its importance, research funding cannot continue to grow at a rate faster than the GNP and the overall growth in the federal budget. Senior officials have privately wondered whether America can afford continued high levels of R&D outlays across the board. Even those who do not seek major organizational changes or a transformation of the research system believe that the time is ripe to think through priorities.

A promising beginning effort is the National Academy of Science's white paper, "Federal Science and Technology Budget Priorities," prepared in response to a request by Congress for "advice on developing an appropriate institutional framework and information base for conducting cross-program development and review of the nation's research and development programs. This should be structured in such a way that it can be used by both the Executive Branch and Congress as a method for reviewing program contents and strategies and for determining funding and organizational priorities for science and technology." [11]

The report divided science and technology activities funded in the federal budget into four nonexclusive categories. The first was those activities relating to the achievement of agency goals and objectives, including R&D on nuclear and alternative energy in the Department of Energy, submarine acoustics in the Defense Department, cell biology in the NIH, and plant disease resistance in the Department of Agriculture. The second category comprised those activities relating to support for scientific personnel, facilities, and knowledge of general use distributed across many different agencies and under the jurisdiction of several congressional committees. They include student fellowships, equipment and instrumentation programs, educational materials and curriculum development for elementary and secondary schools, and facilities to care for experimental animals. The third category covered urgent national priorities of the president and Congress that are likely to involve a variety of executive agencies and congressional committees, including attempts to understand and control global environmental change, AIDS research, work relating to strategic nuclear defense, research on fusion energy, and industrial development research in biotechnology, superconductivity, and manufacturing technologies. The fourth category comprised activities relating to major programs that have significant effects on overall budget levels—the Superconducting Supercollider, the proposed space station, and mapping and sequencing the human genome.

The report concludes that the agencies generally do a good job of deciding which technologies and research programs help them achieve their own program goals. It also suggests that close coupling of R&D funding and agency missions should remain a prominent feature of the research system. Decentralization works well when agency missions are reasonably well defined and the oversight needed is akin to normal OMB review of departmental budget requests. The system works less well for the activities that sustain training, research, and technical infrastructure. These activities are especially apt to be neglected by the mission agencies when budgets are tight, so that the president's science adviser and OMB must seek an overall assessment of their value to the nation.

R&D programs that transcend departmental jurisdictions and contribute to major national priorities are also likely to require special scrutiny. New programs have difficulty in reaching a critical mass of effort, particularly if they do not clearly fit within established agency missions. The NAS does not fully spell out the implications of its analysis, but one infers that the growth in established agency programs, including R&D, may have to be held in check to permit new priorities to be met. The NAS does not explicitly make the argument, but its position also clearly suggests that large facilities such as the Superconducting Supercollider or the human genome project that do not directly advance national priorities should face much greater budgetary hurdles. Left unanswered is how funding trade-offs should be made among facilities that do help meet important national priorities, or how new priorities should be weighed against established agency activities in the event of a Gramm-Rudman-Hollings sequester or other dramatic reduction in federal spending.

The NAS report also suggests changes to improve the congressional process of setting priorities on programs that cross departmental jurisdictions. The House and Senate budget committees and the authorizing committees with broad responsibilities for science and technology policy would evidently conduct reviews of relevant matters before the budget was disaggregated for agency-by-agency and program-by-program examination. The appropriations committees would also consider matters cutting across programs before spending authority was allocated to the respective subcommittees.

This framework is likely to be most useful in making choices on allocations of new R&D resources. The assumption is that there will be incremental expenditure growth or, conceivably, marked improvement in the outlook for investments in R&D.[12]

But an optimistic outlook for R&D funding does not seem to accord with fiscal realities. Dramatic growth is not likely to occur through cuts in domestic social programs or additional constraints on defense expenditures (the elusive "peace dividend"). Without a resolution of the deficit problem, science policymakers will likely experience increasingly hard choices. Cuts in defense spending will, in particular, put pressure on the defense R&D budget. The case will have to be made that modernization programs deserve priority over force readiness—a difficult argument—and advocates of modernization would have to agree that R&D efforts are more important than production and deployment of new systems. Pressure to cut defense spending will almost certainly accentuate the drive for deficit reduction, and the transfer of defense research into already hard-pressed civilian agencies is not likely. Most importantly, the logic of endless growth in R&D budgets has simply collided with the fiscal realities of American politics. Scientists, like any other group, will have to rethink their priorities and show that they are able to accomplish more with fewer resources.

Revitalizing Government Research

The capabilities of the 700 government laboratories spending nearly $20 billion annually have deteriorated. The *Challenger* disaster and the shutdown of the Energy Department's nuclear reprocessing facilities are only the most vivid illustrations of the practical problems that beset the government's technical bureaus. The problem lies partly in the forces eroding the efficiency of the civil service as a whole. The cap on federal pay and the attacks on the civil service in the 1976, 1980, and 1984 presidential campaigns have made government employment less attractive than in the past. But special problems have also affected scientists as a group. The federal government has never adequately tried to create a career ladder for technical employees who do not move quickly, or perhaps want to move, into a management position. Now they are typically blocked at the GS-13 level, eligible only for modest in-grade pay increases at fixed intervals. Rigidities in recruitment and the acquisition of supplies and equipment also pose growing difficulties. The attractiveness of career opportunities outside the government contributes to the problem, though it also benefits the nation as talent migrates to the private sector.

Government needs to develop a career path for scientists and engineers and reward them in a manner commensurate with their ability. It needs to develop something akin to the position of senior scientist in industry or the professorship or endowed chair in the universities. The nation benefits when government, industry, and universities are healthy and technical work of high quality is carried on in each sector. Competition for skills and some movement of talented people from sector to sector is healthy. A chronically weak sector, however, is a drain on the system. The danger is that many government technical organizations, once strong, may fall into such a state.

Demonstration projects allowed under Title VI of the Civil Service Reform Act of 1978 have attempted to confront the problem. The first project was initiated in July 1980 at the Naval Ocean System Center in San Diego and the Naval Weapons Center in China Lake, California.[13] The aim has been to permit government laboratories to hire scientists and engineers at pay scales within broad bands of the regular civil service categories, thus enabling them to pay salaries higher than those usual at entry levels, and to promote deserving workers more rapidly. Their salaries thus become more comparable with those in the private sector. In theory this is to be accomplished without increasing the agencies' budgets. If the experiments prove successful, policies applying to the entire government could follow.

More radical initiatives have sometimes been proposed for particular national laboratories or federally funded R&D centers. In 1978 the OMB proposed converting part of NIH into a private biomedical research university, but the idea ran into a hornet's nest of opposition from biomedical researchers and is now dormant. The concept continues to appeal to those who believe that government should compete with the private sector in delivering services and should fund them through user fees or other self-generated revenue whenever possible. But any wholesale spin-off of governmental laboratories to the private sector is neither feasible nor desirable. Such steps could turn more researchers loose to compete for a relatively fixed pool of research funds and exacerbate the problems facing already hard-pressed research centers struggling to adapt to fiscal constraints and maintain some scope for new investigator-initiated grants. Most of the government's technical activities were initiated to serve specific government needs. The government laboratories can continue to serve the nation by doing their assigned tasks well. If those tasks are no longer needed, laboratories may have to be closed.

In a further attempt to revitalize research Congress passed the Technology Transfer Act of 1986, which makes it easier for industry to commercialize spin-offs from contract research for the government. Scientists in government laboratories are now permitted to receive royalties or other financial incentives from inventions made while in government employment. Industrial scientists have many avenues to be rewarded for commercial successes resulting from their work, and the government laboratories can reasonably experiment in these areas. In principle, mechanisms need to be found to encourage and reward government scientists interested in promoting the transfer of technology that may grow out of their regular government work. Still, there is little encouragement for aiding industry at present, primarily because government scientists fear accusations of seeking personal gain and worry about potential criminal penalties under the plethora of new ethics regulations. The mood of investigative journalism, inspectors general, and aggressive congressional oversight committees ferreting out wrongdoing has made the situation worse. But by interacting with industry through buffer organizations such as trade associations, the government can do much to remove the taint of conflict of interest.

Table 6-3.   **Average Expenditures for 1980 and 1982 and Publications and Rates in 1984 in Two Scientific Disciplines, by Selected OECD Country**
Expenditures in millions of constant 1982 dollars

Field	United States	Japan	Germany	France	United Kingdom	Nether-lands
Astronomy and Astrophysics						
Expenditure	128.5	68.2	81.2	69.3	72.8	30.0
Publications	1,584	112	215	141	375	85
Publications per million dollars of expenditure	12.3	1.6	2.6	2.0	5.1	2.8
Nuclear and particle physics						
Expenditure	450.1	113.2	224.1	297.6	112.1	48.6
Publications	1,242	160	373	181	156	69
Publications per million dollars of expenditure	2.8	1.4	1.7	0.6	1.4	1.4

Source: OECD, *Science and Technology Policy Outlook, 1988* (Paris, 1988), table 5.

Strengthening the Universities

The nation's universities have been a source of great research productivity since World War II. Widely recognized as the world leaders in many scientific fields, American university scientists and engineers have enjoyed extraordinary successes even in the recent period when American industry has suffered from intense foreign competition. By measures such as Nobel Prizes received or publications by their faculties, the success of American research universities continues. Rates for publications per unit of research for two capital-intensive fields of science shows the apparent strength of the research system (table 6-3).

There are troubling questions, however, about the long-term vitality of the university research system. One concern is whether the universities may have compromised their future health by inadequate investment in physical infrastructure. Broader matters include whether the universities have been active enough in addressing the nation's concerns about technology development. A further is whether the United States can continue to be a world leader in science and engineering in the face of striking weaknesses at lower levels of the educational system. Finally, many scientists fear that politics may be undermining the integrity of federal decisions on allocating resources for science and technology.

Recent studies have criticized the neglect of the buildings and equipment on which university research efforts depend.[14] Maintaining this institutional base was not systematically thought about in the early postwar years, but because funding for the expansion and rapid growth of higher education included considerable capital funds, the matter was moot. In those years federal programs, state policies, and philanthropy supported institutions. When the growth of research funding slowed in the late 1960s, however, federal support for R&D plant all but disappeared. While funding for research picked up again in fiscal 1977, there was no concomitant increase for funding plant and equipment. And state legislatures were also cutting back support for university facilities. A crisis in infrastructure has been the predictable result.

Universities have been able to cope partly by borrowing, issuing revenue bonds, and covering costs of maintenance and repair through higher indirect costs on research grants. They have also become much more aware of capital needs in their accounting, planning, and management practices. But much more remains to be done. The federal government, the states, private philanthropy, and the universities themselves all need to assume greater responsibilities for physical infrastructure. The federal government cannot merely buy particular research projects from the universities without concern for the health of the institutions themselves. This point is generally conceded, but the awareness has not typically led to concrete remedies. Even in a time of fiscal constraint, investment needs must be accorded high priority. For this to happen, the universities themselves will have to plan for the future, even at some cost to current research. That such sacrifice will be a very difficult challenge at a time of intense competition for research funds and when administrative staffs have grown more rapidly than the research and teaching staffs hardly requires elaboration.

Beyond federal policy and university action, state governments must clarify their responsibilities for the financing and upkeep of capital facilities. Industry, stimulated in part by recent legislation, has already increased its collaborative research with universities, which has increased funding for university facilities.[15] Traditional philanthropy is less important in supporting facilities than when the scale of expenditures was smaller, but it can continue to be useful: consider the establishment of the Whitehead Institute at MIT, the gift of a $75 million optical telescope to be operated by the California Institute of Technology, and the grants by the Hughes Medical Institute in selected fields, as well as a host of smaller but still significant philanthropic efforts.

The proper balance among types of funding for university R&D activities has also become a problem. The American system has never been limited to supporting only small projects. Rather, it has been a mix of formula, block, and project grants from a variety of public and private sources. But there has always been a question of how to achieve the best balance among the different kinds of funding.

University scientists have been considered especially creative because of the close linkage between teaching and research in American universities, the intense competition for research funds, and the tendency for scientists to develop early in their careers the habits of independent investigators. The U.S. and the British university systems appear generally alike in relying on competitive projects to finance the research of individual scientists, whereas the continental European countries rely more on block funding to research institutes, which may often be separated from or only loosely attached to universities. Project support to a principal investigator cannot, however, by itself provide for a healthy university research system, and although firm judgments are difficult to make, the nation may have relied too heavily on the project grant to individual researchers in recent years. As figure 6-2 shows, project support as a proportion of total NSF funding rose from 50 percent to 70 percent between 1971 and 1979, primarily because of decreasing institutional support from 1970 to 1976 and the phaseout of the Research Applied to National Needs (RANN) program in the late 1970s. Fellowship support has been under pressure since the early 1960s. The NSF's support for its engineering centers has, however, climbed to 8 percent of its outlays.

The current absence of institutional support is striking. From 1960 to 1968 such support accounted for 20 to 25 percent of the NSF research budget each year. This seeming stability was deceptive, however. The growth can be attributed to the rapid buildup in funding for astronomy, large field projects, oceanographic ships, and computing facilities. Support for facilities contracted sharply during the 1969 budget crisis; funding for one program alone—university computing facilities—dropped from $61 million annually to $39 million in two years. Modest increases beginning in 1972 mostly reflected program transfers from the Defense Department, notably the National Magnet Laboratory. The only significant recent increase resulted from the Advanced Computing Center program, begun in 1985. The total absence of institutional support may suggest some of the shortcomings of the present system: too much reliance on projects has been exacerbated by grants for very limited periods, which impose time pressures on investigators, particularly young scientists who must achieve a quick payoff to justify continued support and

Figure 6-2. **Percentage of NSF Funding for Research and Related Activities, by Category, Fiscal Years 1951–87**

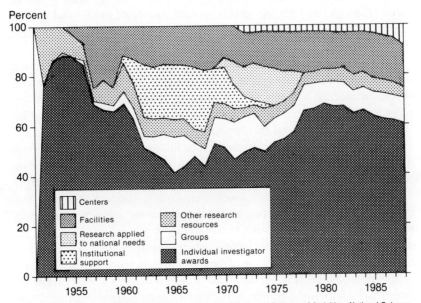

Percent

Legend:
- Centers
- Facilities
- Research applied to national needs
- Institutional support
- Other research resources
- Groups
- Individual investigator awards

Source: National Science Foundation, *Report on Funding Trends and Balance of Activities: National Science Foundation, 1951–1988*, NSF 88-3 (Washington, 1988), p. 8.

to gain tenure. The sheer number of proposals vying for limited resources has led to heightened career tensions among university scientists and anything but a tranquil life. The relatively passive role of agency funding officials who rely too heavily on proposal pressure may not always have served to aid the orderly development of scientific fields. University administrators should also assume more leadership in charting the directions of their institutions' scientific efforts rather than abdicating to the sometimes anarchic behavior of faculty entrepreneurs.

Most importantly, perhaps, the emphasis on projects has focused on quick payoffs and what is required to get the immediate job done, the analogue perhaps of industrial managers' short time horizons. When combined with resource limits, this outlook has reduced incentives for providing fellowships and research assistantships and for modernizing laboratories, precisely the components that nurture the technology base. Agency officials and university administrators need to think of system needs, to balance support for human resources and facilities, coherent research clusters and individual investigators, and to assess areas of comparative advantage at the level of the region, institution, or department.

There is no universal recipe for federal research support agencies to follow. They will vary in their styles of operation, whether they emphasize large grants or a greater number of smaller grants. What is critical is that the needs of the system of higher education and advanced training be considered as well as the capacity of particular universities, departments, or investigators to advance given research goals.

At various times and in various agencies coherent policies have been more nearly realized. Perhaps the NIH in the late 1950s and early 1960s under Director James Shannon was one of the conspicuous successes of reconciling system maintenance needs with efficient use of resources.[16] The NIH was able to combine institutional, block, and project support, along with grants for fellowhips, traineeships, and facilities, to produce the modern health research empire. The system also allowed scientists to choose the direction of their research while it encouraged public and congressional support and understanding. Recently, budget pressures have limited funding for fellowships and traineeships, potentially tilting the system toward shorter-term goals. The NIH has also encountered some difficulties in maintaining support for centers as the demand to provide an adequate number of new investigator-initiated project awards has grown (figure 6-3). Otherwise, the NIH has adapted by cutting the proportion of funding for R&D contracts in half.

At the opposite pole from the individual project is "big science," dependent on large investments in particular scientific instruments or technologies and typified by the superconducting supercollider, the space station, and the human genome project. Such massive efforts pose problems far different from those of small projects: the displacement of individual creative energies into one all-consuming task, the potential crowding out of resources available for other endeavors, and the difficulty of weighing the productivity of the many research teams involved. Big science reflects mission orientation, which emphasizes achieving a few goals at the expense of gradually diffusing technological awareness throughout society. Discoveries in some fields, of course, can only be pursued this way. Means must be found to share costs internationally on such projects, to avoid premature commitment to costly technologies, and to weigh the impact of these efforts on human resources and overall research funding.

Funding is now an acute concern in mapping and sequencing the human genome, in which scientists would seek to determine the order of

Figure 6-3. **Allocation of NIH Extramural Awards, by Category of Activity, Fiscal Years 1979–88**

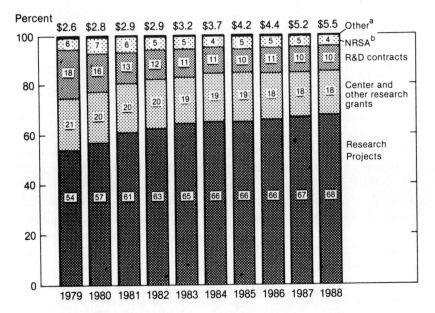

Source: Division of Research Grants, National Institutes of Health, *Extramural Trends, FY 1978–1987* (Washington, 1988), p. 6.
a. Includes construction, medical library grants, and international training grants in epidemiology (fiscal 1988 only).
b. Includes pre-NRSA training.

3 billion base pairs in the DNA.[17] The pursuit is defended on scientific and economic grounds in that success may confer an advantage in information processing and in the techniques essential to maintaining dominance in genetic engineering and biotechnology generally. Critics contend that more numerous teams pursuing different approaches will offer a better chance for scientific advance and for building a strong industrial base in biotechnology and genetic engineering. The debate cannot be resolved here, but of course science does not progress solely through numerous small projects or one megaproject. Mid-level approaches—centers, interdisciplinary institutes, teams—are also important for universities and their patrons in government and industry.

A persistent criticism of America's research universities is that they are closed systems, driven by internal incentives of publication and recognition from peers for activities that bear little or no relation to wider social and economic needs.[18] A contrasting image is that universities are populated with shameless charlatans who blow with the winds of funding fashions and exaggerate the practical applications of their work. There is some truth to both criticisms. Research universities have not sought, or played, an effective role in technology transfer and economic development. (Agriculture is an exception: the land-grant institutions have provided valuable assistance to farmers, but partly because they have emphasized diffusing knowledge rather than generating new knowledge, the agriculturalists have often been second-class citizens in the universities.) While some few exotic technologies have attracted the interest of leading scientists and engineers, it has usually been at a stage far removed from product development, although some research in biomedicine has seemed virtually to merge basic inquiry and product development.

Universities' contribution to economic development can be viewed from two perspectives. As the source of advanced training and research, the university system should lay the basis for future development and help realize the goals of a civilized and learned society. Under this conception, any short-term contribution to economic development would be incidental. If society accepted this paradigm, academic science would not be supported at the levels it aspires to. Society might wish to support academic science generously out of altruism but not on the expectation of immediate contributions to economic growth, job creation, or industrial competitiveness.

The alternative perspective is that the university research system should be more directly engaged in economic development, and that the institutional culture in the universities and federal grant-making agencies should be modified to reflect this perspective. Engaging the universities in achieving short-term goals without explicitly altering incentive structures is, in this view, merely to invite disillusion when the goals are not met. Federal officials should pay more attention to applied research and demonstration projects (the reverse of actual trends since the late 1970s). Universities should encourage and reward scientists and engineers who have interests in applied research. Greater selectivity in supporting basic research is urged. Deborah Shapley and Rustum Roy are vigorous exponents of this view:

> The postwar science system has spread basic research thinly around many institutions and has made university scientists working on applied problems unhappy. A greater concentration of basic science effort may be needed. Perhaps ony the twenty top-ranked universities should remain striving for world-class basic science at the frontier. . . .

We would bring about a value shift . . . to remove the tilt towards basic-science-is-best which has been overlaid on the universities for the past 30 years. For the main middle block of universities would continue some basic research, as overhead for their main work, which would be purposive basic science, applied science, and engineering.[19]

The MIT Commission on Industrial Productivity has also called for changes in the universities so that students would be educated "to be more sensitive to productivity, to practical problems, to teamwork" and has urged that "the federal government's support of research and development should be extended to include a greater emphasis on policies to encourage the downstream phases of product and process engineering."[20]

Events seem to have moved the universities in this direction. They have sought closer collaboration with industry and local officials supervising economic development, partly to sustain research funding and partly because researchers have been attracted to the popular cause of rebuilding America's industrial competitiveness.[21] The National Academy of Sciences has become the National Academy of Sciences-National Academy of Engineering-Institute of Medicine (NAS-NAE-IOM) complex as a symbol of engineers' and physicians' insistence on coequal status with their scientific colleagues.

It is probably unrealistic, however, to expect that research universities will become prominent means of technology transfer, at least not for the myriad smaller companies so important to economic health. For these companies, community colleges are a more likely source of training in computer use, statistical process controls, and management. Certainly, research institutions are unlikely to be prominently involved in near-term product development, even as they move toward refocusing their attention and resources on the applied sciences.

Nor should there be any single pattern for all universities, which can serve their communities, regions, and country in many ways. Local circumstance and decentralized choice will and should continue to shape the evolution of research and teaching efforts. Institutions may increasingly target their scientific opportunities in light of both theoretical and practical concerns, but this should not be disturbing; good science has always involved choice of fields, methodology, and research strategy. University-industry linkages will evolve in various patterns and will blend the broad diffusion of knowledge with the search for new knowledge in equally diverse ways.

Evaluating Peer Review

A persistent matter troubling the government-science relationship has been the status and effectiveness of peer or merit review. The loss of congressional confidence in the process has apparently led to the pork barrel amendments for funding scientific facilities and construction that have raised fears that the whole structure of objective judgment of scientific proposals could be undermined. Even some scientists have grown discouraged and have quietly urged their colleagues to begin considering alternatives to peer review. But despite imperfections, the process remains indispensable to the nation's research system. Particularly when resources are limited, the idea of a fair and objective basis for allocating funds and producing the most efficient outcomes deserves affirmation.

About 80 percent of all citations in the scientific literature are accounted for by 15 percent of the publishing scientists (similarly, 90 percent of all scholarly citations appear in 10 percent of the journals). Science is, in this sense, an elitist activity. Ways must be found, therefore, to see that those who can use resources most productively get them. But this cannot be done in a fashion that incorporates a privileged status for any institution or principal investigator. The United States is precluded, by virtue of its political culture, from seriously considering some alternative funding mechanisms that make sense in other countries. The senior scientists of the Max Planck Society, for example, receive and in turn allocate a sizable fraction of the Federal Republic of Germany's basic science expenditures, a system virtually impossible to replicate in the United States. This is not to say that block funding is unknown here, but it depends on (and may be indirectly tied to) the investigator's performance in peer review competition. Review of individual proposals is the only practicable way, in short, to reconcile the elite character of science with the norms of American democracy. The system is, furthermore, a self-correcting and evolving procedure, not a static and rigid formality.

Peer review dates back at least to the seventeenth century, when the Royal Society of London established a board of editors to evaluate reports submitted for publication in its *Proceedings*.[22] Peer review procedures for federally supported research in the United States were initiated in 1902, when the Fifty- seventh Congress established a scientific advisory board of nongovernmental scientists to assist the surgeon general in the administration of the Hygienic Laboratory (renamed the National Institute of Health in 1930). In 1937 the National Cancer Act created a legal basis for awarding grants through an advisory council; these procedures were extended to grants and fellowships in all health research when the Public Health Service Act was passed in 1944. Two years later, the NIH director created the Division of Research Grants and the initial review groups, which considered proposals and passed them on with ratings and recommendations for approval or rejection—essentially the form of review still in use by the NIH.[23]

The NSF peer review system was developed in the 1950s and has often been modified. The essentials of the system are set out in NSF Manual 10, subsection 122, and NSF Manual 1, subsections 310-90. Under procedures adopted by the National Science Board at its 188th meeting in 1977 and amended at its 251st meeting in 1984, the NSF is required to submit an annual report to the board on its use of peer review during the preceding year, including recommendations for change or reconsideration of the foundation's policies. Other agencies making research grants use variations on these systems, sometimes broadening the responsibilities of agency program officials and relying on fewer formalities in soliciting external advice.

The number of studies, examinations, and external and internal reviews of the NIH and NSF review policies is astonishing. In addition to the annual NSF study submitted to NSF Director Erich Bloch in 1989 and the eighteen-month study completed by the NIH in 1988, various congressional committees, external review committees mandated by Congress, and independent studies have examined the operation and administration of peer review, the recurring issues, and the strengths and weaknesses of the system.[24] While some important changes have resulted from their recommendations, they have found the system generally sound.

The traditional criticisms of peer review are that the system is unfair, is dominated by an old boy network, and displays favoritism; the paperwork burden is excessive and the process too time-consuming; and there is an inherent bias against high-risk proposals, including those that do not fall neatly within established disciplinary categories. Studies do not support the contention of an old boy network or other systematic favoritism. Membership on the peer panels has become more diverse. For example, the proportion of women on NIH panels increased from 19 percent to 21 percent between 1977 and 1987, the proportion of physicians declined slightly, and representation of minority investigators grew from 5.7 percent to 16 percent.[25] The average age of NIH study section members, approximately 45, has remained relatively constant over the years; in 1987 only 19 percent of section members were older than 50. The

Table 6-4.   **Attitudes of Principal Investigators toward Fairness of National Science Foundation Peer Review System, by Outcome of Applications**
Percent

Applications outcome	Satisfied	Neutral	Dissatisfied	Percentage of applicants surveyed
Refused consistently	27	16	57	28
Refused frequently	40	16	44	14
Refused once	36	21	44	13
Accepted consistently	83	5	12	13
Accepted frequently	61	11	28	26
Accepted once	87	7	6	5

Source: National Science Foundation, 1988

distribution of section members geographically has remained proportional to the research efforts in the various regions of the country (which is also the case for the NSF).

For both NIH and NSF awards, competition has grown more intense. Scores necessary to achieve awards have been higher, fewer awards relative to the total number of applications have been made, and service on review panels or study sections does not appear to have a marked effect on success ratios.[26] For NSF awards, state, region, or affiliation with a prestige university appears to have little effect on success ratios. Renewal applications do fare better in some NSF divisions than others, but new female principal investigators are treated equally with new male principal investigators. About 11 percent of 135,000 science and engineering faculty members in the United States apply to the NSF each year (13 percent of chemists, 10 percent of mathematicians, 4 percent of economists, 20 percent of earth scientists, 14 percent of electrical engineers, 16 percent of computer scientists), and the foundation rejects approximately two-thirds of the applications.[27]

As an indication of the competitive nature of the overall research support system, while the top twenty universities from 1967 to 1984 continued to receive 42 percent of total federal research funds (down from 45 percent), four new universities were represented in the top twenty in 1984.[28]

The sense of fairness with which university researchers view the peer review system is difficult to assess precisely, but in a recent NSF survey of 14,200 principal investigators, 62 percent of those who responded were satisfied with the process, or were neutral; 38 percent were dissatisfied. Not surprisingly, table 6-4 shows that the more successful researchers tended to view the system in more favorable terms. Perhaps surprisingly, however, 56 percent of those frequently rejected and 57 percent of one-time rejects were either satisfied or neutral. Even 43 percent of those frequently turned down were either satisfied or neutral.

Both the NIH and NSF have been conscious of the paperwork burden of the peer review process and have taken steps to alleviate the number of submissions and streamline the process while maintaining fair and orderly procedures. In fiscal 1988 the NSF received 24,161 proposals involving 140,994 requests for written reviews (of which 114,312 were actually received). The NSF has some 170,000 reviewers in divisional files, of which 55,999 were asked in 1988 to serve either as ad hoc reviewers or as members of standing panels; one-third of the reviewers turn over annually.[29] The NIH's Division of Research Grants had 68 chartered study sections, some with subcommittees meeting separately, for a total of 92 groups with approximately 1,500 reviewers in fiscal 1987; an additional 40 review groups existed under the program heading of Bureaus, Institutes, Divisions. In total they awarded an estimated 21,000 research grants of all kinds to continuing and new centers and projects in fiscal 1988.[30] Each year from 1980 to 1987 the NIH awarded new competing grants, from 4,785 in fiscal 1980 to 6,446 in fiscal 1987.[31]

To address this considerable burden of work, the NIH and the NSF recommended greater use of electronic submissions and improved agency computer data bases and less boilerplate in applications. Each has instituted small prizes and special awards for young scientific investigators.

The NIH decided that extending the duration of grants would be a major improvement because it would allow more time for research and reduce the number of proposals that would have to be submitted and reviewed each year. In 1988 the average length of award had increased to 3.92 years from 3.15 years in the late 1970s (figure 6-4). Programs developed to extend grant duration include the Javits Award of the National Institute of Neurological Disorders and Stroke and the Outstanding Investigator Grant of the National Cancer Institute.

In mid-1988 the NSF's Engineering Division initiated a new program of expedited awards for novel research that eliminated external review and sought to stimulate creative, innovative work. As of February 1989, 239 grants had been awarded. The program was limited to one-time grants of up to $30,000; total awards were not to exceed 5 percent of the division's research budget. A similar foundation mechanism encour-

Figure 6-4. **Average Duration of Project for National Institutes of Health RO1 Awards, with Range of Individual Bid Averages, Fiscal Years 1977–88**

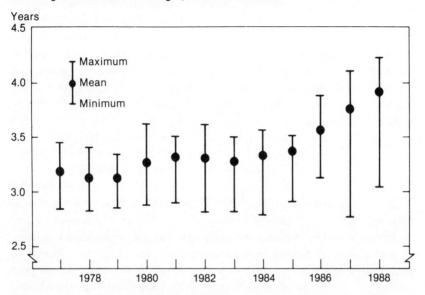

Source: National Institutes of Health, *Report of the NIH Peer Review Committee, 1988* (Bethesda, Md., 1988), figure 10.

ages grants of up to $50,000 for exploratory work and is again limited to 5 percent of the program's budget.

The criticism that peer review leads to conservatism has given evaluators the most difficulty. The exceptional investigator awards, the new scientist programs, and the expedited reviews represent administrative actions to combat potential conservative bias, but officials concede that there is no perfect solution. The exploratory grants, however, partly meet the need to encourage ventures into emerging fields, support research crossing traditional program boundaries, and help experienced researchers shift fields. Multidisciplinary research, especially, requires care in the review procedures and is necessarily more difficult to evaluate than traditional research. Despite the problems of identifying the peers for the review, the merits of proposals continue to receive informed, objective weighing.

In spite of its flaws, the peer review process as it has evolved in the American research support system has no substitute. It is flexible and can accommodate new concepts, and it provides the sense of due process that makes the unequal allocation of resources acceptable in a democratic society. The logic and the basic precepts of the American research system depend on the evaluation and support of individual projects. Coherent area grants or awards to centers are also dependent on projects. No scientific investigator typically depends exclusively on the block grant for support. Rather, there is a rough proportionality between the additional research funds investigators get through peer-reviewed projects and the amounts they may receive from the block grant at the host institution or center. Even institutional grants partly depend on competition, often pegged to the amounts received in individual project competition or by the institution on the basis of student enrollment choices. As for grants to large facilities, peer review no longer operates easily. But even here scientists have tempered the operation of pork barrel considerations by making certain that sites selected have technical merit as well as political support.

The peer review system is the only rational way to allocate resources in a pluralistic endeavor for which resources are always insufficient to satisfy all claimants. As resources are stretched, the criticisms of peer review will inevitably mount, but the neutral appraisal of proposals and the efficient use of resources will be all the more urgent. Avoiding large mistakes will have a higher priority than ever. No peer review system can be perfect, and no system can ever fully escape the tendency to follow prevailing orthodoxies. The answer to that problem is for research to continue to rely on multiple sources of support.

Modernizing American Industry

Modernizing its industry is widely perceived as one of America's most pressing challenges. Few doubt the importance of maintaining—or re-covering—leadership in the technologies with important applications for commercial products and processes. The chief difficulty lies in isolating issues involving technology from the context of public and private actions that contribute to the nation's competitiveness. Yet several factors seem clear. The United States has lagged behind competitors in the percentage of GNP devoted to investment. From 1971 to 1980 the U.S. rate was 6.6 percent, while the United Kingdom's was 8.1 percent, Italy's 10.7 percent, West Germany's 11.8 percent, France's 12.2 percent, and Japan's 19.5 percent.[32] However, in the 1980s productivity growth increased for the business sector as a whole, reflecting efforts to achieve higher product quality, and is seen most clearly in increased manufacturing productiv-

Table 6-5. **Annual Growth of Manufacturing Labor Productivity, by Country, Selected Periods, 1960–88**

Country	1960–73	1973–79	1979–88
United States	3.2	1.4	3.3
Canada	4.5	2.1	2.2
Japan	10.3	5.5	5.8
France	6.4	4.6	3.1
Germany	5.8	4.3	2.6
Italy	6.4	5.7	4.1
United Kingdom	4.2	1.2	4.7
Belgium	6.9	6.0	5.4
Denmark	6.4	4.2	1.1
Netherlands	7.4	5.5	3.5
Norway	4.3	2.2	2.6
Sweden	6.4	2.6	3.0

Source: U.S. Bureau of Labor Statistics, 1989.

ity.[33] The increase in business R&D expenditures in the late 1970s and early 1980s may have encouraged the growth in productivity. But other nations have not stood still. Indeed, Japan, Italy, the United Kingdom, Belgium, and the Netherlands have exceeded the U.S. productivity growth rate from 1979 to 1988, and Sweden and France have nearly matched it (table 6-5).

The strength of the U.S. economy is beyond dispute. American firms continue their lead in technological advances in computers, biotechnology, pharmaceuticals, jet aircraft, scientific instruments, and organic chemicals. But loss of technological leadership in key sectors could bring sudden and dramatic reversals in the nation's fortunes. The ability to commercialize technology or to achieve mass markets for new products after succeeding in the early stages of innovation is the Achille's heel.

In the United States . . . outstanding successes in basic science and in defense research have left the product-realization process a poor cousin. . . . American companies evidently find it difficult to design simple, reliable mass-producible products; they often fail to pay enough attention at the design stage to the likely quality of the manufactured product; their product development times are excessively long; they pay insufficient attention to manufacturing processes; they take a reactive rather than a preventive approach to problem solving; and they tend to underexploit the potential for continuous improvement in products and processes.[34]

Table 6-6.   **Percentage of U.S. Plants That Have Adopted Programmable Automation, by Industry, 1989**

Industry	Percentage
Guided missiles	100
Engines and turbines	89
Aircraft and parts	56
Metal forging and stamping	35
Construction and related activities	35
Motor vehicles and equipment	26
Heating equipment and plumbing fixtures	15
Farm and garden machinery	9
Jewelry, silverware, and plateware	2
Average	44

Source: Mary Ellen Kelley and Harvey Brooks, "The State of Computerized Automation in U.S. Manufacturing," John F. Kennedy School of Government, Harvard University, 1989.

Moreover, the technology needed for the nation to become more fully competitive is spread thinly. Industrial plants in many sectors have hardly begun to use programmable automation in their manufacturing processes, and only 44 percent of sample plants have adopted computer-aided design or manufacturing techniques (table 6-6).

It is evident that potentially far-reaching changes and dislocations are occurring worldwide. The shifts may or may not produce fewer jobs, slower growth, or lower living standards for the United States. New economic patterns of cooperation and competition are emerging, for which even an adequate analytical vocabulary is lacking. New global capital flows will influence economic behavior in yet uncharted ways. Economic insecurity is rife. Yet Americans, with their capacity for self-renewal and their restless energy, seem well positioned to prosper.

There is no magic potion to guarantee competitiveness. Technological leadership in industry and the universities is critical, but so are the capacities to manufacture, distribute, and market high-quality goods and services, to finance and manage this marketing effectively, and to get the most out of human resources at all levels. Myriad actions of the household, school, firm, and government have contributed to the problems of productivity and competitiveness, and only broad efforts in many quarters can solve them.

For the individual firm, the quality of the design team or the amount of automation in the production process is not an end in itself. They are the means to meet the company's objectives. For the nation the same is true. A modernized industrial base is not an end in itself; it is a means to a healthy economy serving the ends of a free society. The nation's energies should be directed toward improving all aspects of technology diffusion and economic performance, not simply improving engineering or manufacturing operations.

Breaking the Regulatory Gridlock

The policy problems posed by the shift in public concern from obvious pollution to more uncertain and invisible threats to human health and the environment are easy to state but difficult to solve. Ideally, scientific research, including the quantitative techniques of risk assessment, should be able to identify the most serious hazards, taking into account the interrelated effects of pollutants on different parts of the ecosystem. Then the experts should be able to devise solutions for neutralizing them. As further evidence becomes available, appropriate modifications would be made so that improvement is constant. With due deference to the experts, public officials would support this experimental approach, balancing the need for reasonable margins of safety with the need to avoid unnecessary regulatory burdens on industry. Indeed, President Richard M. Nixon called for such an approach in a message accompanying his 1970 reorganization of federal environmental responsibilities, declaring that an "effective approach to pollution control would identify pollutants; track them through the entire ecological chain, observing and recording changes in form as they occur; determine the total exposure of man and his environment; examine interactions among forms of pollution; [and] identify where on the ecological chain interdiction would be most appropriate." [35]

Unhappily, this ideal bears little resemblance to actual environmental politics. There is scarcely any neutral matter at any stage—from identifying environmental problems to assessing scientific testing results to specifying solutions. Even if one knew the ideal solution to a particular problem, merely stating it would not guarantee that the solution could be adopted. The conflicting interests, mobilized over two decades of controversy, are alert to any effort to alter the agenda.

A legacy persists of environmental statutes adopted on the assumption that weak enforcement and lack of will were the problems. Although the statutes contain numerous unenforceable provisions and unattainable objectives, attempts to change the framework of environmental law run up against formidable obstacles. Disputes that cannot be resolved in the political process are transferred into the courts, which are ill-equipped to handle complex technical issues. Overall the situation is marked by highly mobilized group activism, bold rhetoric but cautious action, and battles over symbols and talismanic signs; often the ends of both effective regulation and sensible reform are defeated.

Is there a way out of this impasse? There are grounds for cautious optimism if one believes ideas have the power to alter institutions. In the past several decades risk assessment has made notable progress in sorting out the benefits and costs of environmental policies.[36] The quality of science applied to environmental problems has improved, and new disciplines have emerged to study them. Greater awareness of environmental issues has diffused among the public, creating continued strong support for environmental goals along with greater realism as to what can be achieved and how soon. Improved use of science can help to break the regulatory gridlock over protecting the environment, provided that Americans can live with the ambiguity of a democratic polity in which opposing sides rarely score major triumphs but seldom suffer total defeat. The environmental movement, initially exhibiting the excessive faith in technological fixes that it deplored in its enemies, has gradually become more realistic in its assessment of the ill effects of technology, in its faith in the capacities of scientists to devise cost-free solutions to environmental problems, and in its dealings with industry. Industry, for its part, not only tolerates but needs regulations to protect the environment, and asks only that the framework be stable.

The research system has already made great strides in solving such relatively easy problems as removing tangible pollutants from air and water. Many of the acute risks posed by accident or catastrophe have also been brought under control. A recent assessment concludes that "the United States has done much better at avoiding health and safety catastrophes than most people realize, considering the vast scope and magnitude of the threats posed by the innovativeness of science and industry in the twentieth century. . . . Recognizing and appreciating the strengths of catastrophe-aversion systems may give us the inspiration to envision the next steps."[37] Modest successes include the nation's system for ensuring the safety of nuclear power plants (with the remaining need for off-site storage of spent fuel), provisions for the safe handling and transportation of dangerous chemicals, improved worker safety, the creation of a stable regulatory regime for the commercial development of biotechnology, and the beginnings of multilateral action to protect the ozone layer and combat acid rain.

There are some partial successes even for chronic, long-term environmental and health dangers. After a decade of warnings from the surgeon general and mounting evidence linking smoking to cancer, heart disease, and other health hazards, Americans are smoking less. They have also changed their dietary habits to limit the consumption of cholesterol and fats and have become aware of the benefits of physical fitness. The Chemical Manufacturers Association has proposed for the first time to set operating and safety standards that its 170 member firms would have to meet to retain membership.[38]

Problems not solved and requiring greater attention include pollution in homes and office buildings from asbestos, for instance, or radon, the chronic long-term health risks posed by various toxic substances, the acute air quality problems and noncompliance with clean air standards in some cities, and the complex issues associated with global warming trends. The trade-offs between the necessary exploitation of energy sources and the ideal of a clean environment will continue to pose problems, and sharp disagreements between the industrialized countries and third world nations preoccupied with economic development will require great technical and diplomatic efforts. For these and other critical environmental issues the contributions of science in monitoring trends and identifying critical points at which intervention can produce measurable results will be indispensable. Targets have to be identified before they can be met, and science will help pinpoint the new issues.

In September 1988 the Science Advisory Panel of the EPA called for a greater R&D effort in the agency and more emphasis on preventing environmental degradation.[39] This is the route toward sound progress. Policies to protect the environment, so often marked by controversy and tumult, nevertheless seem to become more effective. The growing recognition of the need to address the most important problems, as identified by research not emotion, is slowly moving policy away from the deadlocks of the past. A democratic polity is the stronger for the gradual seepage of ideas through numerous layers of opinion as the nation makes up its mind on an important problem.

Conclusions

This book has traced the twists and turns of American science and technology policy since the end of the Second World War. In three phases the nation struggled to understand the role of new scientific and technological forces in public life. Assumptions about the goals of policy often

changed, yet the debates seemed always to retrace familiar lines of argument and to display continuities amid the changes.

Thus at the core of this inquiry lies a paradox: how is it that one can
detect major changes in the postwar period as a whole, while at any given
time science policies seem to change very little? This paradox can be
explained if change in these policies is viewed as a gradual process of
responding to or anticipating specific problems. There has been both
more activity and less than meets the eye at any given time. More activity
because even when matters appear settled, the forces of change have
worked to alter the substratum of ideas and erode the existing framework of policy. And less because what has seemed like a dramatic departure has often been merely a reconfiguration of familiar elements.

The driving force has been the power of ideas. Vannevar Bush's or
Harley Kilgore's conception of what was good for the nation set the initial terms of the debate, and their allies and descendants have carried on
the argument. The fight was not primarily about power or short-run interests, though these were certainly present, but rather about the kinds
of policies that would best promote scientific and technological progress
and about the essential purpose of support for science programs—to
disseminate existing knowledge or to ensure a steady supply of new
knowledge? This clash of values lay at the heart of science policymaking.
And from the beliefs and aspirations of those involved were distilled the
ideas for specific policies and institutional arrangements. The substance
of the ideas was a critical, though sometimes neglected, component. The
point is summed up well by John Kingdon: "Political scientists are accustomed to such concepts as power, influence, pressure, and strategy. If
we try to understand public policy solely in terms of these concepts, however, we miss a great deal. The content of the ideas themselves, far from
being mere smokescreens or rationalization, are integral parts of decisionmaking." [40]

Studies of health, disability, energy, regulatory, and unemployment
and relief policies, showing how ideas gradually diffuse through the political system and shape the agendas of action, have come to similar conclusions. [41] Far-reaching changes are wrought in the framework of social
action, but only after lengthy preparation as proposals are debated, discussed, and refined. The politicians, lobbyists, and bureaucrats on whom
political scientists have traditionally focused their attention are neither
removed from the debate nor relegated to minor roles. They are vigorous
participants in the politics of ideas.

The alchemy by which participants in democratic politics transmute

the often inchoate desires of millions of people into a more or less working reality while sustaining the confidence that each person has some
stake in the process is more admirable when the daunting nature of the
task is recognized. Politicians are still the gatekeepers for policy ideas.
That they are sometimes overwhelmed and fail to produce orderly results
only emphasizes the contribution they make to democracy when they do
their job well.

Great events at first glance appear to be missing from this account of
gradual change. If revolutions occur in science itself, would it not be
fruitful to look for revolutionary changes in science policy? The story
told here is not wholly at odds with the idea of dramatic changes. The
consensus on science policy grew out of the profound transformation
wrought by a world war. In turn the unraveling of the beliefs and attitudes constituting the consensus was accelerated by another war and its
wrenching social consequences.

Because of the long lead times in many programs and because of the
constituencies supporting current policies, however, parts of the old system have remained in place, and interested parties continue to appeal to
the old beliefs. But a narrower perspective has replaced system values.
Individual survival has become the dominant motive, and in consequence
policy has become more fragmented. Science affairs have moved from
being a cluster of concerns with a center of gravity to a looser collection
of themes. Science policy has increasingly overlapped with other matters
on the national agenda. In one sense science policy has seemed less coherent because it has become a more important, more integral part of
the most urgent national concerns, for which there are no strictly technical solutions but which have technical aspects inseparable from broad
policy considerations.

There can be no new consensus in science policy. America now seeks
to maintain a mature system, not to build anew. The crisis facing the
nation is less dramatic than it was after World War II, but more complex.
Yet the concern for the design of the research system shown by the founders of science policy is very badly needed today. We must seek a greater
coherence and rationality in the policies the nation pursues to nurture
science and technology and in the myriad ways that scientific resources
are brought to bear on national problems. But in the end, of course, the
controlling vision that the nation needs relates not only to science and
technology. A more inclusive vision must place science and technology
within the context of the basic directions of public life and confront the
next steps in the adventure of American democracy.

Notes

Chapter One

1. See John Arthur Passmore, ed., *Priestley's Writings on Philosophy, Science and Politics* (Collier Books, 1965).

2. My periodization is close to, but differs slightly from, those of such other scholars as Harvey Brooks, "National Science Policy and Technological Innovation," in Ralph Landau and Nathan Rosenberg, eds., *The Positive Sum Strategy: Harnessing Technology for Economic Growth* (National Academy Press, 1986), pp. 128–35; Brooks, "What Is the National Agenda for Science and How Did It Come About?" *American Scientist*, vol. 75 (September–October 1987), pp. 513–17; David Dickson, *The New Politics of Science* (Pantheon, 1984); and *A History of Science Policy in the United States, 1940–1985*, Committee Print, House Committee on Science and Technology, 99 Cong. 2 sess. (Government Printing Office, 1986).

3. Harvey Brooks, "The Scientific Advisor," in Robert Gilpin and Christopher Wright, eds., *Scientists and National Policy Making* (Columbia University Press, 1964), pp. 73–96.

4. Wallace S. Sayre, "Scientists and American Science Policy," in Robert Gilpin and Christopher Wright, eds., *Scientists and National Policy Making* (Columbia University Press, 1964), pp. 97–112.

5. Jurgen Schmandt and James Everett Katz, "The Scientific State: A Theory with Hypotheses," *Science, Technology and Human Values*, vol. 11 (Winter 1986), pp. 40–52.

6. Don K. Price, *America's Unwritten Constitution* (Louisiana State University Press, 1983). Also relevant are Price, *The Scientific Estate* (Harvard University Press, 1965), and Price, *Government and Science: Their Dynamic Relation in American Democracy* (New York University Press, 1954).

7. For a broad interpretation see Price, *Scientific Estate*, chap. 5. See also Harvey A. Averch, *A Strategic Analysis of Science and Technology Policy* (Johns Hopkins Press, 1985); W. Henry Lambright, *Presidential Management of Science and Technology: The Johnson Presidency* (University of Texas Press, 1985); Landau and Rosenberg, eds., *Positive Sum Strategy*; Bernard Barber, *Science and the Social Order* (Free Press, 1952); James L. McCamy, *Science and Public Administration* (University of Alabama Press, 1960); and A. Hunter Dupree, *Science in*

the Federal Government: A History of Policies and Activities to 1940 (Harvard University Press, 1957).

8. Vannevar Bush, *Science: The Endless Frontier* (National Science Foundation, 1945, reprinted 1960).

9. See Harold D. Lasswell and Daniel Lerner, eds., *The Policy Sciences: Recent Developments in Scope and Method* (Stanford University Press, 1951).

10. Michael Polanyi, "The Republic of Science: Its Political and Economic Theory," *Minerva*, vol. 1 (Autumn 1962), p. 62.

11. James Sterling Young, *The Washington Community, 1800–1828* (Columbia University Press, 1966).

12. Alvin M. Weinberg, "Criteria for Scientific Choice II: The Two Cultures," *Minerva*, vol. 3 (Autumn 1964), p. 12.

13. Alvin M. Weinberg, "Values in Science: Unity as a Criterion for Scientific Choice," *Minerva*, vol. 22 (Spring 1984), pp. 1–12.

14. Donald Kennedy, "Government Policies and the Cost of Doing Research," *Science*, February 1, 1985, pp. 480–81.

15. Robert K. Merton, "Priorities in Scientific Discovery," in Bernard Barber and Walter Hirsch, eds., *The Sociology of Science* (Free Press of Glencoe, 1962), pp. 447–85.

16. Price, *Scientific Estate*, p. 171.

Chapter Two

1. *Transactions of the American Philosophical Society*, vol. 1 (1771), p. xvii, quoted in John C. Greene, *American Science in the Age of Jefferson* (Iowa State University Press, 1984), p. 6.

2. *Memoirs of the American Academy of Arts and Sciences*, vol. 1 (1785), p. vii, quoted in Greene, *American Science*, p. 6.

3. George H. Daniels, *Science in American Society: A Social History* (Knopf, 1971), chap. 7.

4. Richard Harrison Shryock, "American Indifference to Basic Science during the Nineteenth Century," *Archives Internationales d'Histoire des Sciences*, vol. 5 (1948), pp. 50–65, reprinted in Walter Hirsch and Bernard Barber, eds., *The Sociology of Science* (Free Press, 1978); and Nathan Reingold, "American Indifference to Basic Research: A Reappraisal," in George H. Daniels, ed., *Nineteenth Century American Science: A Reappraisal* (Northwestern University Press, 1972), pp. 38–62.

5. Robert V. Bruce, *The Launching of Modern American Science, 1846–1876* (Knopf, 1987).

6. Greene, *American Science*, pp. 6–7.

7. Bruce, *Launching of Modern American Science*, chaps. 13, 14; and I. Bernard Cohen, "Science and the Civil War," *Technology Review*, vol. 48 (January 1946), pp. 167–70, 192, 193.

8. In this case Bache acted without the assistance of his friend Joseph Henry,

who opposed the idea of the Academy. Bruce, *Launching of Modern American Science*, pp. 223–24, 301–05.

9. See, for example, Greene's analysis of the developments in astronomy and chemistry in *American Science*, pp. 128–57, 160–75.

10. Bruce, *Launching of Modern American Science*, chap. 6; and Daniels, *American Science in the Age of Jackson* (Columbia University Press, 1968).

11. Tocqueville's pronouncement that Americans were practical rather than theoretical was based on his observation of American businessmen, tradesmen, and other people in practical walks of life. There is no evidence that he observed or exchanged views with scientists. See Alexis de Tocqueville, *Democracy in America* (Vintage Books, 1945), vol. 2, pp. 46–47; and Bruce, *Launching of Modern American Science*, chap. 9.

12. Bruce, *Launching of Modern American Science*, pp. 313–25.

13. Dael Wolfle, *The Home of Science: The Role of the University* (McGraw Hill, 1972), pp. 45–51; and Bruce, *Launching of Modern American Science*, pp. 326–38.

14. Roger L. Geiger, *To Advance Knowledge: The Growth of American Research Universities, 1900–1940* (Oxford University Press, 1986).

15. See, however, A. Hunter Dupree, *Science in the Federal Government; A History of Policies and Activities to 1940*, (Harvard University Press, 1957), pp. 149–83; John M. Gaus and Leon O. Wolcott, *Public Administration and the United States Department of Agriculture* (Chicago: R. R. Donnelley, 1940); Vernon W. Ruttan, "Technical Change and Innovation in Agriculture," in Ralph Landau and Nathan Rosenberg, eds., *The Positive Sum Strategy: Harnessing Technology for Economic Growth* (National Academy Press, 1986), pp. 333–54 (and the references cited on pp. 355–56); and Jean Mayer and Andre Mayer, "Agriculture, the Island Empire," *Daedalus*, vol. 103 (Summer 1974), pp. 83–95.

16. *Report of the Advisory Committee to the U.S. Department of Agriculture* (National Academy of Sciences, 1973). For a discussion, see Bruce L. R. Smith and Joseph J. Karlesky, *The State of Academic Science*, vol. 1: *Summary of Major Findings* (New Rochelle, N.Y.: Change Magazine Press, 1977), pp. 136–38.

17. Jacob Bigelow, *Elements of Technology, Taken Chiefly from a Course of Lectures Delivered at Cambridge on the Application of Science to the Industrial Arts* (Boston: Hilliard, Gray, Little, and Wilkins, 1831).

18. Paul Mantoux, *The Industrial Revolution in the Eighteenth Century: An Outline of the Beginnings of the Modern Factory System in England* (Harper and Row, 1961); E. J. Hobsbawn, *Industry and Empire: An Economic History of Britain since 1750* (Letchworth, England: Garden City Press, 1968), especially pp. 40–60; and David S. Landes, *The Unbound Prometheus: Technological Change and Industrial Development in Western Europe from 1750 to the Present* (Cambridge University Press, 1969), especially pp. 41–123.

19. Robert Luther Thompson, *Wiring a Continent: The History of the Telegraph Industry in the United States, 1832–1866* (Princeton University Press, 1947), pp. 20–24.

20. Don K. Price, *Government and Science: Their Dynamic Relation in American Democracy* (New York University Press, 1954), chaps. 1, 3; and see, generally, David A. Hounshell, *From the American System to Mass Production, 1800–1932: The Development of Manufacturing Technology in the United States* (Johns Hopkins University Press, 1984); and Alfred D. Chandler, Jr., *The Visible Hand: The Managerial Revolution in American Business* (Harvard University Press, 1954).

21. Hounshell, *From the American System to Mass Production*, pp. 43–61.

22. Frank W. Taussig, *The Tariff History of the United States*, 6th ed. (Putnam's, 1914), pp. 128–35.

23. Alfred D. Chandler, Jr., and Richard S. Tedlow, *The Coming of Managerial Capitalism: A Casebook in the History of American Economic Institutions* (Richard D. Irwin, 1985), pp. 327–30; and Chandler, *Visible Hand*, pp. 290–99.

24. Chandler, *Visible Hand*, pp. 484–500.

25. Wickham Skinner, "The Taming of the Lions: How Manufacturing Leadership Evolved, 1780–1984," in Kim B. Clark, Robert H. Hayes, and Christopher Lorenz, eds., *The Uneasy Alliance: Managing the Productivity-Technology Dilemma* (Harvard Business School Press, 1985), pp. 63–100.

26. On the rise of industrial research, see Leonard S. Reich, *The Making of American Industrial Research: Science and Business at G.E. and Bell 1876–1926* (Cambridge University Press, 1985); George Wise, *Willis R. Whitney, General Electric, and the Origins of U.S. Industrial Research* (Columbia University Press, 1985); Neil H. Wasserman, *From Invention to Innovation: Long-Distance Telephone Transmission at the Turn of The Century* (Johns Hopkins University Press, 1985); A. Hunter Dupree, *Science in the Federal Government* (Harvard University Press, 1957), pp. 287–88, 305–23; and David A. Hounshell and John Kenly Smith, Jr., *Science and Corporate Strategy: DuPont R&D, 1902–1980* (Cambridge University Press, 1988). See also Margaret B. W. Graham, "Corporate Research and Development: The Latest Transformation," *Technology in Society*, vol. 7 (1985), pp. 179–95.

27. For a time the Lazzeroni of Alexander Bache and Joseph Henry and a small handful of colleagues exercised a measure of leadership. But as the system grew large and more complex it became more decentralized. Dupree, *Science in the Federal Government*, pp. 115–19.

28. Dupree, *Science in the Federal Government*, pp. 215–31.

29. The term *regulatory state* is borrowed from Chalmers A. Johnson, who uses it to contrast American adversarial relations between business and government since World War II with the Japanese partnership arrangements. See *MITI and the Japanese Miracle: The Growth of Industrial Policy, 1925–1975* (Stanford University Press, 1982). See also Alfred D. Chandler, Jr., "Government versus Business: An American Phenomenon," in John T. Dunlop, ed., *Business and Public Policy* (Harvard University Press, 1980), pp. 1–11; and Bruce L. R. Smith, "The Adversarial Nature of Government-Business Relations in the United States," in Karen A. Szymanski, ed., *The Donald S. MacNaughton Symposium*

Proceedings: U.S. Competitiveness in the World Economy (Syracuse University Press, 1983) pp. 137–51.

30. Much of the information in this section is from Dupree, *Science in the Federal Government*, pp. 302–40.

31. Peter J. Kuznick, *Beyond the Laboratory: Scientists as Political Activists in 1930s America* (University of Chicago Press, 1987), p. 10.

32. Geiger, *To Advance Knowledge*, pp. 103–05.

33. Quoted in Dupree, *Science in the Federal Government*, p. 341. See also Herbert Hoover, "The Nation and Science," *Science*, vol. 65 (1927), pp. 26–29.

34. Quoted in Dupree, *Science in the Federal Government*, p. 342. For an elaboration, see J. McKean Cattell, "Scientific Research in the United States," *Science*, February 12, 1926, p. 188.

35. Kuznick, *Beyond the Laboratory*, pp. 18–41.

36. Charles Wiener, "Physics in the Great Depression," *Physics Today*, vol. 23 (October 1970), pp. 31–36.

37. Quoted in Dupree, *Science in the Federal Government*, p. 354. See also K. T. Compton, "Report of the Science Advisory Board," *Science*, vol. 81 (January 1935), p. 15; and Kuznick, *Beyond the Laboratory*, pp. 51–64.

38. Memorandum to the Secretary of the Interior, February 12, 1935, quoted in Dupree, *Science in the Federal Government*, p. 357. See also Karl T. Compton, "Put Science to Work," *Technology Review*, vol. 37 (1935), p. 133.

39. Lewis E. Auerbach, "Scientists in the New Deal: A Pre-War Episode in the Relations between Science and Government in the United States," *Minerva*, vol. 3 (Summer 1965), pp. 457–82. See also Donald S. Swain, "The Rise of a Research Empire: NIH, 1930 to 1950," *Science*, vol. 138, p. 3323.

40. Carroll W. Pursell, Jr., "Legislation and the National Bureau of Standards," *Technology and Culture*, vol. 9 (April 1968), p. 163. See also Pursell, "Government and Technology in the Great Depression," *Technology and Culture*, vol. 20 (January 1979), pp. 162–74.

41. R&D expenditures and personnel increased at an average of 7 percent a year.

42. Robert Jungk, *Brighter Than a Thousand Suns: A Personal History of the Atomic Scientists* (Harcourt Brace Jovanovich, 1958); Daniel J. Kevles, *The Physicists: The History of a Scientific Community in Modern America* (Knopf, 1978); and Richard G. Hewlett and Jack M. Holl, *Atoms for Peace and War, 1953–61* (University of California Press, 1989).

43. Irvin Stewart, *Organizing Scientific Research for War: The Administrative History of the Office of Scientific Research and Development* (Little Brown, 1948).

Chapter Three

1. This failure to recognize scale effects is evident, for example, in the early history of DDT. On the importance of nonlinearities with respect to scale, see Harvey Brooks, "The Typology of Surprise in Technology, Institutions, and De-

velopment," in W. C. Clark and R. E. Munn, eds., *Sustainable Development of the Biosphere* (Cambridge University Press, 1986), pp. 325–47. For specific examples see William M. Stigliani, "Changes in Valued Capacities of Soil and Sediments as Indicators of Non-Linear and Time-Delayed Environmental Effects," *Environmental Monitoring and Assessment*, vol. 10 (1988), pp. 245–307.

2. National Science Foundation, *National Patterns of Science and Technology Resources* (Washington, 1984), p. 2.

3. President's Scientific Research Board, *Science and Public Policy: Administration for Research*, 3 vols. (Government Printing Office, 1947), vol. 1, p. 26. (Hereafter referred to as the Steelman report.)

4. Organization for Economic Cooperation and Development, *Science, Growth, and Society: A New Perspective* (Paris, 1971).

5. Harvey Brooks, "Future Needs for the Support of Basic Research," in *Basic Research and National Goals: A Report to the Committee on Science and Astronautics, U.S. House of Representatives* (Washington: National Academy of Sciences, March 1965), pp. 74–110.

6. Daniel J. Kevles, "The National Science Foundation and the Debate over Postwar Research Policy, 1942–1945," *Isis*, vol. 68 (1977), p. 6. See also the very useful collection of excerpts from congressional hearings and critical reports in James L. Penick, Jr., and others, eds., *The Politics of American Science, 1939 to the Present*, rev. ed. (MIT Press, 1972); and Irvin Stewart, *Organizing Scientific Research for War: The Administrative History of the Office of Scientific Research and Development* (Little, Brown, 1948).

7. Quoted in Kevles, "National Science Foundation," p. 8.

8. Kevles, "National Science Foundation," p. 12. The idea for this report originated with Oliver Carr, a wartime presidential adviser, who thought it would be used to support the president in the 1944 reelection campaign. Bush objected to this timetable and to the partisan overtones, and devised the questions he would respond to and the timetable he would follow. I am indebted to Oscar Reubhausen, who handled the negotiations between Bush and Carr (acting on President Roosevelt's behalf), for sharing his recollections of these events.

9. Vannevar Bush, *Science, the Endless Frontier* (Washington: National Science Foundation, 1945, reprinted 1960), p. 12. Subsequent references are to the 1960 edition. Excerpts from the Bush report appear in Penick and others, *Politics of American Science*, pp. 106–15.

10. Bush, *Science*, p. 13.

11. President's Science Advisory Committee, *Improving the Availability of Scientific and Technical Information in the United States* (White House, 1958); President's Science Advisory Committee, *Scientific Progress, the Universities and the Federal Government* (GPO, 1960); and David Bell, *Report to the President on Government Contracting for Research and Development* (White House, 1962), reprinted in *Systems Development and Management*, Hearings before a House Subcommittee of the Committee on Government Operations, 87 Cong. 2 sess. (GPO, 1962), pt. 1, pp. 191–337; and National Academy of Sciences, *Basic*

Research and National Goals, A Report to the Committee on Science and Astronautics, U.S. House of Representatives (Washington, 1965).

12. Bush, *Science*, pp. 95, 96.

13. Bush, *Science*, p. 94.

14. National Research Council, "Research in Europe and the United States," *Outlook on Science and Technology: The Next Five Years* (W. H. Freeman, 1982), chap. 13.

15. Bush, *Science*, p. 21.

16. W. J. Lescure, ed., *Forty Years of Excellence* (Washington: Office of Naval Research, 1986); and Penick and others, *Politics of American Science*, pp. 22–24.

17. Nick A. Komons, *Science and the Air Force: A History of the Air Force Office of Scientific Research* (Arlington, Va.: Office of Army Research, 1966); Harvey Sapolsky, "American Science and the Military: The Years since the Second World War," in Nathan Reingold, ed., *Sciences in the American Context: New Perspectives* (Smithsonian Press, 1979); Robert Sigethy, "Air Force Organization for Basic Research, 1945–1970: A Study in Change," Ph.D. dissertation, American University, 1980; and Dorothy Nelkin, *Science and the Military* (Cornell University Press, 1972).

18. National Academy of Sciences, *Basic Research and National Goals*, p. 23; and Penick and others, *Politics of American Science*, pp. 33–34. The rationale for the role of the mission agencies was articulated in the Eisenhower administration in the context of an internal debate over the responsibilities of NSF as compared with those of the mission agencies. In executive order 10521, March 17, 1954, President Eisenhower said, "As now or hereafter authorized or permitted by law, the Foundation shall be increasingly responsible for providing support by the Federal Government for general-purpose basic research through contracts and grants. The conduct and support by other Federal agencies of basic research in areas which are closely related to their missions is recognized as important and desirable especially in response to current national needs, and shall continue." Quoted in Penick and others, *Politics of American Science*, p. 34.

19. Alan Waterman, "Introduction," in Bush, *Science*, p. xxv.

20. Harvey Brooks, "What Is the National Science Agenda, and How Did It Come About?" *American Scientist* (September–October 1987), p. 513.

21. Hugh Heclo and Lester Salamon, eds., *The Illusion of the Presidency* (Boulder, Colo.: Westview Press, 1981), p. 26.

22. Penick and others, *Politics of American Science*, pp. 33–34, for the concept of general purpose basic research, and National Academy of Sciences, *Basic Research and National Goals*, p. 12, for the idea of the balance wheel. Curiously the balance wheel role was not articulated until the NSF began to be a larger and more central presence.

23. Even with the expansion of its budget in recent years, the NSF's share of the federal R&D budget is only slightly more than 3 percent.

24. Bush, *Science*, p. 28.

25. Steelman report, vol. 3, pp. 23, 29.

26. Bruce L. R. Smith, "Accountability and Independence in the Contract State," in Bruce L. R. Smith and Douglas C. Hague, eds., *The Dilemma of Accountability in Modern Government: Independence vs. Control* (St. Martin's Press, 1971), pp. 3–69.

27. Morris Janowitz, *The Professional Soldier: A Social and Political Portrait* (Free Press, 1974), chap. 2.

28. This is known as the 6.1 function, based on defense budget category 6, research and development. The other functions are 6.2, exploratory development; 6.3, advanced development; 6.4, engineering development; 6.5, management and support; and 6.6, operations systems development. Kosta Tsipis and Penny Janeway, eds. *Review of U.S. Military Research and Development* (Washington: Pergamon-Brassey's, 1984), p. 7.

29. For an authoritative description of source selection, see J. Ronald Fox, *Arming America: How the U.S. Buys Weapons* (Harvard University Press, 1974), especially pp. 15–25, 259–86.

30. Herbert F. York, *Making Weapons, Talking Peace: A Physicist's Odyssey from Hiroshima to Geneva* (Basic Books, 1987). See also John Logsdon, *Decision to Go to the Moon: Apollo Project and the National Interest* (University of Chicago Press, 1976).

31. Bruce L. R. Smith, *The Rand Corporation* (Harvard University Press, 1966).

32. Bush, *Science*, p. 20.

33. The most recent review is Office of Science and Technology Policy, *Report of the White House Science Council: Federal Laboratories Panel* (1983); and OSTP, *Progress Report on Implementing the Recommendations of the White House Science Council's Federal Laboratory Review Panel*, vol. 1: *Summary Report*, and vol. 2: *Status Reports by Agencies* (1984).

34. See Smith, *Rand Corporation*, pp. 16–18; and Bruce L. R. Smith, "The Future of the Not-For-Profit Research Corporation," *Public Interest*, no. 8 (Summer 1967), pp. 127–42.

35. For aircraft development see David C. Mowrey and Nathan Rosenberg, "The Commercial Aircraft Industry," in Richard R. Nelson, ed., *Government and Technical Progress: A Cross-Industry Analysis* (Pergamon, 1982), pp. 101–61; and Robert Schlaifer and S. D. Heron, *The Development of Aircraft Engines and Fuels* (Harvard Graduate School of Business Administration, 1950). For computers see Kenneth Flamm, *Targeting the Computer* (Brookings, 1988); Flamm, *Creating the Computer* (Brookings, 1988); and Flamm and Thomas L. McNaugher, "Rationalizing Technology Investments," in John D. Steinbrunner, ed., *Restructuring American Foreign Policy* (Brookings, 1988), pp. 119–84.

36. Although the government was the major customer, "development [of engines] was fully successful only when the services gave the firms all possible freedom in deciding on details of design and development. Government intervention in technical details led to very considerable delay, and often to a poorer product

in the end," Schlaifer and Heron, *Development of Aircraft Engines and Fuels,* p. 8.

37. Bush, *Science,* p. 21.

38. Steelman report, vol. 1, p. 4.

39. Harvey A. Averch, *A Systematic Analysis of Science and Technology Policy* (Johns Hopkins University Press, 1985), pp. 12–16, 43–47.

40. Bush, *Science,* p. 21.

41. In 1981 industry won a 25 percent tax credit for R&D expenditure beyond the previous year's level as part of the Reagan administration's Economic Recovery Tax Act. The credit survived, in somewhat reduced form, the tax reform of 1986. See Joseph A. Pechman, *Federal Tax Policy,* 5th ed. (Brookings, 1987), pp. 137–54.

42. For example, in 1960 the highest tax bracket for ordinary income was 90 percent. Capital gains stood at 25 percent, leaving a 65 percent spread; in 1976 ordinary income was taxed at a maximum of 50 percent, capital gains at 42 percent; in 1986, 27 percent and 20 percent. The inference is sometimes drawn that the 1986 tax reform was a step in the wrong direction. Managers, under this reasoning, may tend to act less like owners and more like employees who have little stake in the long-term appreciation of the company's stock. The leveraged buyouts of the late 1980s may, however, reflect a step back toward the fusion of ownership and management. On capital gains, other economists dispute the contention that differential tax treatment encourages investment and assert that tax reform of the kind enacted in 1986 provides encouragement for capital formation. See Pechman, *Federal Tax Policy,* pp. 116–22, for an analysis of the capital gains issue. See also Congressional Budget Office, *How Capital Gains Tax Rates Affect Revenues: The Historical Evidence* (1988).

43. Bush, *Science,* p. 21.

44. Congressional Budget Office, *Using Federal R and D to Promote Commercial Innovation* (1988), pp. 52–53, 64–66.

45. For a lucid analysis of postwar economic thinking, see Herbert Stein, *Presidential Economics* (Simon and Shuster, 1984).

46. See Stein, *Presidential Economics.*

47. Harold Green and Alan Rosenthal, *Government of the Atom: The Integration of Powers* (New York: Atherton Press, 1963).

48. National Research Council, *Regulating Pesticides in Food: The Delaney Paradox* (Washington: National Academy Press, 1987), pp. 25–26.

49. Shawn Bernstein, "The Rise of Air Pollution as a National Political Issue: A Study of Issue Development," Ph.D. dissertation, Columbia University, 1982.

50. See Detlev Bronk, "Science Advice to the White House: The Genesis of the President's Science Advisors and the National Science Foundation," in William Golden, ed., *Science Advice to the President* (Pergamon, 1980), pp. 254–56.

51. Bernard Bailyn and Donald Fleming, eds., *The Intellectual Migration: Europe and America 1930–1960* (Harvard University Press, 1969).

52. William Diebold, "The End of ITO," Department of Economics and the Social Institute, Princeton University, October 1952; and William Adams Brown, *The U.S. and the Restoration of World Trade: An Analysis and Appraisal of the ITO Charter and the General Agreement on Tariffs and Trade* (Brookings, 1950).

53. Wickham Skinner, "The Taming of the Lions: How Manufacturing Leadership Evolved, 1780-1984," in Kim B. Clark, Robert H. Hayes, and Christopher Lorenz, eds., *The Uneasy Alliance: Managing the Productivity Technology Dilemma* (Harvard Business School Press, 1985) pp. 63-100.

54. Organization for Economic Cooperation and Development, *Science, Growth, and Society* (Paris, 1971).

55. Jean-Jacques Servan-Schreiber, *The American Challenge* (Atheneum, 1968). See also Bruce L. R. Smith, "A New 'Technology Gap' in Europe?" *SAIS Review*, vol. 6 (Winter-Spring, 1986) pp. 219-36.

56. Robert Gilpin, *France in the Age of the Scientific State* (Princeton University Press, 1968).

57. Raymond Vernon, "International Investment and International Trade in the Product Cycle," *Quarterly Journal of Economics*, vol. 80 (May 1966), pp. 190-208.

58. Robert Gilpin, *The Political Economy of International Relations* (Princeton University Press, 1987).

59. Bush, *Science*, p. 29.

60. National Research Council, *Balancing the National Interest: U.S. National Security Export Controls and Global Economic Competition* (Washington: National Academy Press, 1987), pp. 71-73.

61. See U.S. Congress, Senate Committee on Foreign Relations, *Atoms for Peace Manual*, S.Doc 55 (GPO, 1955); Arnold Kramish, *The Peaceful Atom in Foreign Policy* (Harper and Row, 1963); and Phillip Mullenbach, *Civilian Nuclear Power: Economic Issues and Policy Format* (New York: Twentieth Century Fund, 1963).

62. Louis Hartz, *The Liberal Tradition in America* (Harcourt Brace, 1955).

63. See Alfred P. Sloan, Jr., *My Years at General Motors* (Doubleday, 1964).

64. The phrase is attributed variously to publisher Henry Luce and to the wartime president of the U.S. Chamber of Commerce. See Peter Collier and David Horowitz, *The Rockefellers: An American Dynasty* (Holt, Rinehart, and Winston, 1976), pp. 226-27; and Stephen S. Rosenfeld, "The American Century Still?" *Washington Post*, August 5, 1988, p. A23.

Chapter Four

1. Harvey A. Averch, *A Strategic Analysis of Science and Technology Policy* (Johns Hopkins University Press, 1985), pp. 16-20, 56-59.

2. John T. Wilson, *Academic Science, Higher Education and the Federal Government, 1950-1983* (University of Chicago Press, 1983), pp. 30-42.

3. Quoted in Bruce L. R. Smith and Joseph J. Karlesky, *The State of Aca-*

demic Science: The Universities in the Nation's Research Effort, vol. 1: *Summary of Major Findings* (New Rochelle, N.Y.: Change Magazine Press, 1977), p. 34.

4. National Board on Graduate Education, *Federal Policy Alternatives toward Graduate Education* (Washington: National Academy of Sciences, 1974) chaps. 2, 6; and David E. Drew, *Strengthening Academic Science* (Praeger, 1985).

5. Chalmers W. Sherwin and Raymond Isenson, *First Interim Report on Project Hindsight* (Office of the Director of Defense Research and Engineering, 1966).

6. Stanley Karnow, *Vietnam: The History* (Viking Press, 1983), p. 500; and "McNamara's Fence against the Reds: Will It Really Work?" *U.S. News and World Report*, September 18, 1967, p. 8.

7. The golden fleece awards began in 1975 when the NSF was cited for a study of agression in monkeys. From then until his retirement at the end of the One-hundreth Congress, Proxmire made one award a month, mostly to federal funding agencies for what he regarded as egregiously silly projects.

8. For an analysis of the antiscience currents of the 1970s, see Don K. Price, *The Scientific Estate* (Harvard University Press, 1965), pp. 82–119; Price, "Money and Influence: The Links of Science to Public Policy," *Daedalus*, vol. 103 (Summer 1974), pp. 97–113; Edward Shils, "Faith, Utility, and Legitimacy of Science," *Daedalus*, vol. 103 (Summer 1974), pp. 1–15; Harvey Brooks, "The Technology of Zero Growth," *Daedalus*, vol. 102 (Fall 1973), pp. 139–52; and Leo Marx, *Machine in the Garden: Technology and the Pastoral Idea in America* (Oxford University Press, 1964). Barry Commoner, *The Closing Circle: Nature, Man, and Technology* (Knopf, 1971), is perhaps the most extreme attack on technology in the environmental movement of the 1970s. Jacques Ellul, *The Technological Society* (Random House, 1967), is a classic European antitechnology statement that has some overtones in the American context.

9. See Richard R. Nelson, *The Moon and the Ghetto* (Norton, 1977); and Wallace S. Sayre and Bruce L. R. Smith, *Government, Technology and Social Problems* (Columbia University, Institute for Science in Human Affairs, 1969).

10. Sayre and Smith, *Government, Technology and Social Problems*, pp. 13 and following. On the general mood of the period, see Peter Szanton, *Not Well Advised* (New York: Russell Sage Foundation, 1981).

11. Richard R. Nelson, Merton J. Peck, and Edward D. Kalachek, *Technology, Economic Growth, and Public Policy: A Rand Corporation and Brookings Institution Study* (Brookings, 1967).

12. Alice M. Rivlin, *Systematic Thinking for Social Action* (Brookings, 1971).

13. See Charles J. Hitch and Roland McKean, *The Economics of Defense in the Nuclear Age* (Atheneum, 1965), for a pioneering statement of the role of PPB and systems analysis in the defense policy process.

14. Wilson, *Academic Science*, p. 31. See also House Subcommittee on Science, Research and Development of the Committee on Science and Aeronautics, *Geographic Distribution of Federal Research and Development Funds*, Government and Science Report no. 4, 88 Cong. 2 sess. (1964), p. 48.

15. Quoted in Wilson, *Academic Science,* p. 32. See also memorandum from President Lyndon B. Johnson to heads of departments and agencies, "Strengthening Academic Capabilities for Science through the Country," September 13, 1965; and Drew, *Strengthening Academic Science.*

16. Wilson, *Academic Science,* p. 33. See also the National Science Foundation Act of 1950, as amended, P.L. 90-407 (1968).

17. Wilson, *Academic Science,* pp. 57–59.

18. Smith and Karlesky, *State of Academic Science,* vol. 1, p. 19.

19. The panel noted that "the percentage of extramural funds spent for such research grants between fiscal years 1967 and 1974 has declined from about 60 percent to about 45 percent. The Panel is concerned about this trend. . . . The investigator-initiated research grant has been the primary instrument without which major advances in improving the nation's health would not have been possible." Public Health Service, *Report of the President's Biomedical Research Panel,* DHEW (OS) 76-500 (U.S. Department of Health, Education, and Welfare, 1976), p. 17.

20. Rodney W. Nichols, "Mission-Oriented R&D," *Science,* April 2, 1971, p. 29.

21. House Subcommittee on Science, Research and Technology of the Committee on Science and Technology, *1976 National Science Foundation Authorization,* H.R. 94-930, 94 Cong. 1 sess. (1975), pp. 28–29.

22. Nichols, "Mission-Oriented R&D," p. 29.

23. Harvey Brooks and Roland W. Schmitt, "Current Science and Technology Policy Issues: Two Perspectives," George Washington University, 1985, p. 23.

24. Smith and Karlesky, *State of Academic Science,* vol. 1, pp. 19, 200–05.

25. National Board on Graduate Education, *Federal Policy Alternatives,* chap. 6; Smith and Karlesky, *State of Academic Science,* vol. 1, pp. 161–68; and David W. Breneman, "Effects of Recent Trends in Graduate Education on University Research Capability in Physics, Chemistry and Mathematics," in Smith and Karlesky, *State of Academic Science,* vol. 2: *Background Papers,* pp. 133–62.

26. Smith and Karlesky, *State of Academic Science,* vol. 1, pp. 86–101, especially table 3.

27. Smith and Karlesky, *State of Academic Science,* vol. 1, pp. 54–55.

28. Brooks and Schmitt, "Current Science and Technology Policy Issues," pp. 20–21.

29. Donald Swain, "The Rise of a Research Empire: NIH, 1930 to 1950," *Science,* December 14, 1962, pp. 233 and following. The National Institutes of Health together became roughly three times the size of NSF, a reversal of the pattern that developed in Europe. In France and Germany the health research agencies are a third the size of the agencies supporting general science. The Medical Research Council in the United Kingdom spends less than a third of the research funds given out by the research councils. National Research Council,

Outlook on Science and Technology: The Next Five Years (W. H. Freeman, 1982), chap. 13.

30. Smith and Karlesky, *State of Academic Science,* vol. 1, pp. 24–25.

31. Harvey Brooks, "What Is the National Agenda for Science, and How Did It Come About?" *American Scientist,* vol. 75 (September–October 1987), p. 512.

32. Averch, *Strategic Analysis,* pp. 16–20, 56–59.

33. Richard R. Nelson, *High-Technology Policies: A Five Nation Comparison* (Washington: American Enterprise Institute for Public Policy Research, 1984); and Dorothy Nelkin, *The Politics of Housing Innovation: The Fate of the Civilian Industrial Technology Program* (Cornell University Press, 1971).

34. Averch, *Strategic Analysis,* pp. 59–60.

35. James D. Carroll, "Science and the City," *Science,* February 28, 1969, p. 902.

36. Averch, *Strategic Analysis,* p. 59.

37. Department of Commerce, *Technological Innovation: Its Environment and Management* (1967), pp. 52–54.

38. Averch, *Strategic Analysis,* p. 61.

39. "Operation Breakthrough," *Newsweek,* July 21, 1969, pp. 78–81; "Operation Breakthrough: Commitment and Questions," *Architectural Record,* September 1969, p. 10; and "Romney Soups Up His New Vehicle," *Business Week,* September, 13, 1969, p. 82.

40. Averch, *Strategic Analysis,* pp. 62–63.

41. *A History of Science Policy in the United States, 1940–1985,* Committee Print, House Committee on Science and Technology, 99 Cong. 2 sess. (September 1986), p. 102.

42. See "Transcript of President Nixon's Television Address," *New York Times,* November 8, 1973, p. 32, for discussion of Project Independence.

43. Averch, *Strategic Analysis,* pp. 63–65.

44. Frank Press, "Science and Technology in the White House, 1977 to 1980: Part 1," *Science,* January 9, 1981, pp. 139–45; Press, "Science and Technology in the White House, 1977 to 1980: Part 2," *Science,* January 16, 1981, pp. 249–56; and Averch, *Strategic Analysis,* pp. 65–68.

45. Press, "Science and Technology, Part 1," p. 143.

46. Press, "Science and Technology, Part 1," pp. 141, 144.

47. The most significant recent measure is H.R. 2164, introduced by Representative George Brown in April 1987, to establish a Department of Science and Technology. Congress has been more interested in centralizing science organization in the executive branch than recent presidents have been.

48. Amitai Etzioni, *An Immodest Agenda: Rebuilding America Before the Twenty-First Century* (McGraw-Hill, 1983), pp. 316–17.

49. Allen V. Kneese and Charles L. Schultze, *Pollution, Prices and Public Policy* (Brookings, 1975), chaps. 1–3.

50. *History of Science Policy,* pp. 64–68.

51. Quoted in Dorothy Nelkin, *Science in the Streets* (Priority Press, 1984), p. 28.

52. William R. Manchester, *The Glory and the Dream: A Narrative History of America, 1932-72* (Little, Brown, 1974), pp. 1255-58.

53. Peter Collier and David Horowitz, *The Rockefellers: An American Dynasty* (Holt, Rinehart, and Winston, 1976), pp. 385-88.

54. On the rise of environmental issues and their absorption into the public agenda, see John W. Kingdon, *Agendas, Alternatives, and Public Policies* (Little, Brown, 1984). A lively debate was carried on in the early 1970s over whether local political structures would block effective decisionmaking on environmental issues. See Mathew A. Crenson, *The Unpolitics of Air Pollution: A Study of Non-Decisionmaking in the Cities* (Johns Hopkins Press, 1971). The progressive view held that environmental issues could only be tackled effectively at the national level with uniform standards. During the Reagan administration, preoccupation with national solutions came to be regarded as a disservice to effective policymaking. See William D. Ruckelshaus, "Risk, Science, and Democracy," *Issues in Science and Technology*, vol. 1 (Spring 1985), pp. 19-38.

55. James Q. Wilson, *Political Organizations* (Basic Books, 1974).

56. Kneese and Shultze, *Pollution, Prices, and Public Policy*, pp. 31-32.

57. Kneese and Shultze, *Pollution, Prices, and Public Policy*, chap. 5; Lester B. Lave and Gilbert S. Omenn, *Clearing the Air: Reforming the Clean Air Act* (Brookings, 1981); R. Shep Melnick, *Regulation and the Courts: The Case of the Clean Air Act* (Brookings, 1983); and David M. O'Brien, *What Process Is Due? Courts and Science Policy Disputes* (New York: Russell Sage Foundation, 1987), especially chap. 4.

58. William J. Abernathy, Kim B. Clark, and Alan M. Kantrow, "A New Industrial Competition," *Harvard Business Review* (September–October 1981), pp. 68-81.

59. International Trade Administration, *An Assessment of U.S. Competitiveness in High Technology Industries* (U.S. Department of Commerce, 1983) pp. 41, 53.

60. Harvey Brooks, "Technology as a Factor in U.S. Competitiveness," in Bruce R. Scott and George C. Lodge, eds., *U.S. Competitiveness in the World Economy* (Harvard Business School Press, 1985), p. 335.

61. National Science Board, *Science and Engineering Indicators, 1987* (Washington, 1987), figure 0-26.

62. Harvey Brooks, "National Science Policy and Technological Innovation," in Ralph Landau and Nathan Rosenberg, eds., *The Positive Sum Strategy: Harnessing Technology for Economic Growth* (Washington: National Academy Press, 1986) p. 143.

63. Robert Z. Lawrence, "Changes in U.S. Industrial Structure: The Role of Global Forces, Secular Trends, and Transitory Cycles," in Federal Reserve Board of Kansas City, *Symposium on Industrial Change and Public Policy* (1983), p. 47, quoting a Japanese government 1981 white paper on international trade.

64. National Science Board, *Science Indicators, 1982* (Washington, 1983), pp. 5, 6, 8.

65. National Science Board, *Science Indicators, 1983* (Washington, 1984), p. 13.

66. Brooks, "National Science Policy and Technological Innovation," p. 142.

67. See Committee on Science, Engineering, and Public Policy, National Academy of Sciences, *Balancing the National Interest: U.S. National Security Export Controls and Global Economic Competition* (Washington: National Academy Press, 1987), chap. 4.

68. E. F. Schumacher, *Small is Beautiful: Economics as If People Mattered* (Harper and Row, 1975).

69. For the evolution of science policy in the State Department, see House Committee on International Relations, *Science, Technology, and American Diplomacy: An Extended Study of the Interactions of Science and Technology with U.S. Foreign Policy*, Committee Print, House Subcommittee on International Security and Scientific Affairs, 3 vols. (GPO, 1977), pp. 1909–2107; and Eugene B. Skolnikoff, *Science, Technology and American Foreign Policy* (MIT Press, 1967). I am grateful to David Z. Beckler for sharing his recollections of the Berkner Committee and the origins of the State Department scientific office. The late Frank Huddle of the Congressional Research Service, who compiled the three-volume study, talked to me at length about the history of the department's involvement with science and technology, and the reasons why Congress sought a larger role for the State Department in coordinating the international scientific activities of the U.S. government. I later served as director of policy assessment in what became the State Department's Bureau of Oceans, International Environmental, and Scientific Affairs.

70. Quoted in *Science, Technology, and American Diplomacy*, p. 1358.

71. *Science, Technology, and American Diplomacy*, p. 1369.

Chapter Five

1. Harvey Brooks and Roland W. Schmitt, "Current Science and Technology Policy Issues: Two Perspectives," George Washington University, 1985, pp. 17–18.

2. Intersociety Working Group, *AAAS Report XIII, Research and Development, FY 1989* (Washington: American Association for the Advancement of Science, 1988), pp. 23–27.

3. Simon Ramo, *America's Technology Slip* (John Wiley, 1980).

4. See David Dickson, *The New Politics of Science* (Pantheon, 1984), pp. 38–39. Friedman indicated in "An Open Letter on Grants," *Newsweek*, May 18, 1981, that he favored "major cuts in NSF grants as a step toward the abolition of the NSF."

5. Quoted in Dickson, *New Politics of Science*, p. 39.

6. The first full-year Reagan request was for fiscal 1982. See Willis H. Shapley, Albert H. Teich, and Gail J. Breslow, *Research and Development, AAAS*

Report VI, FY 1982: New Directions for R and D: Federal Budget—FY 1982: Industry, Defense (Washington: American Association for the Advancement of Science, 1981), p. 5. Also see *Special Analyses, Budget of the United States Government, Fiscal Year 1989.* For a commentary on the initial Reagan budget, see Bruce L. R. Smith and James D. Carroll, "Reagan and the New Deal: Repeal or Replay?" *PS*, vol. 14 (Fall 1981), pp. 758–66.

7. Don K. Price, *Government and Science: Their Dynamic Relation in American Democracy* (Oxford University Press, 1962), chap. 2.

8. Detlev Bronk, "Science Advice in the White House: The Genesis of the President's Science Advisors and the National Science Foundation," in William T. Golden, ed., *Science Advice to the President* (Pergamon Press, 1980), pp. 245–56.

9. Bronk, "Science Advice," p. 250.

10. Recollections of David Z. Beckler, executive director to the Science Advisory Committee, at a forum on science and the economy sponsored by the Carnegie Corporation at the University of California, San Diego, in 1988. Beckler served as the executive director of the committee in the Eisenhower administration and continued when the committee was transformed into the President's Science Advisory Committee in 1957. He served until President Nixon abolished the committee in 1973.

11. Harvey A. Averch, *A Strategic Analysis of Science and Technology Policy* (Johns Hopkins University Press, 1985), pp. 12–16.

12. This story was told by Carey at the Carnegie forum on science and the economy at the University of California, San Diego, in 1988.

13. Conversations with William T. Golden and I. I. Rabi.

14. Research and related activities inceased to $69.29 million in fiscal 1959, science and engineering education to $61.29 million, and the U.S.-Antarctic program to $2.31 million. National Science Foundation, *Report on Federal Funding Trends and Balance of Activities: National Science Foundation, 1951–1958,* NSF 88-3 (Washington, December 1987).

15. John T. Wilson, *Academic Science, Higher Education, and the Federal Government, 1950–1983* (University of Chicago Press, 1983), pp. 43–50.

16. Golden, ed., *Science Advice to the President* ; and Golden, ed. *Science and Technology Advice to the President, Congress, and Judiciary* (Pergamon Press, 1988), and the review essay by Bruce L. R. Smith, "Odd Men Out," *Sciences*, vol. 28 (September–October 1988), pp. 44–50.

17. *A History of Science Policy in the United States, 1940–1985,* Committee Print, House Committee on Science and Technology, 99 Cong. 2 sess. (September 1988), p. 98.

18. Golden, ed., *Science and Technology Advice to the President.*

19. *History of Science Policy,* Committee Print, p. 102.

20. Smith, "Odd Men Out."

21. *History of a Science Policy,* pp. 102–03.

22. Claude E. Barfield, "Science Report: Congress Moves to Reset Priorities in Federal Research and Development," *National Journal,* September 30, 1972,

pp. 1524–33. I am indebted to Ellis Mottur for sharing his recollections of these events.

23. *History of Science Policy*, Committee Print, p. 105. G. Richard Allison, formerly Vice President Rockefeller's chief of staff, shared his recollections of the events leading up to the reestablishment of the post of science adviser and the decision to ask Stever to assume the job. Stever agreed to serve until the end of President Ford's term and then step down whatever the outcome of the 1976 election.

24. Averch, *Strategic Analysis*, pp. 65–68; and Frank Press, "Science and Technology in the White House: 1977–1980," parts I and II, *Science*, January 9 and 16, 1981, pp. 139–45, 249–56.

25. Bruce L. R. Smith and Joseph J. Karlesky, *The State of Academic Science: The Universities in the Nation's Research Effort*, 2 vols. (New Rochelle, N.Y.: Change Magazine Press, 1977).

26. Smith and Karlesky, *State of Academic Science*, vol. 1: *Summary of Major Findings*, pp. 168–69. See also Robert M. Rosenzweig, *The Research Universities and Their Patrons* (University of California Press, 1982), chap. 5.

27. National Science Foundation, *Five Year Outlook: Promises, Opportunities, and Constraints in Science and Technology*, vol. 2 (Washington, 1980), p. 275.

28. National Science Board, *Science and Engineering Indicators, 1987* (Washington: National Science Foundation, 1987), pp. 104–05, 295–96.

29. National Science Board, *Science Indicators, 1980* (Washington: National Science Foundation, 1980), pp. 9, 54, 92, 129; and see the discussion in Harvey Brooks, "Technology as a Factor in U.S. Competitiveness," in Bruce R. Scott and George C. Lodge, eds., *U.S. Competitiveness in the World Economy* (Harvard Business School Press, 1985), pp. 347–48.

30. Harvey Brooks, "What Is the National Agenda for Science?" *American Scientist*, vol. 75 (September–October 1987), p. 516. See also Wickham Skinner, "The Taming of Lions: How Manufacturing Leadership Evolved, 1780–1984," in Kim B. Clark and others, *The Uneasy Alliance: Managing the Productivity and Technology Dilemma* (Harvard Business School Press, 1985), pp. 63–100.

31. Initial funding for the program was $1.8 million, but because of enthusiastic support from universities and industry, it rapidly grew. Realization of the need to upgrade mechanical engineering coincided with industry's recognition of a long-standing neglect of manufacturing process technologies. I am indebted to conversations with Ali Seireg, chairman of the ASME committee that recommended the action.

32. See Kenneth Flamm, *Creating the Computer* (Brookings, 1988), chap. 2.

33. Office of technology Assessment, *Commercial Biotechnology: An International Analysis*, OTA-BA-218 (January 1984); and Office of Technology Assessment, *Technology Initiatives and Regional Economic Development: Census for State Government Initiatives for High Technology Industrial Development* (May 1983).

34. The growth of industry-university joint research activity was stimulated

by the Small Business Patent Act of 1980 (and the amendments adopted in 1984 when Congress extended the life of the original act) and the Small Business Innovation Development Act of 1982; see Bruce L. R. Smith, ed., *The State of Graduate Education* (Brookings, 1985), pp. 23–28.

35. William J. Abernathy and Kim B. Clark, *Industrial Renaissance: Producing a Competitive Future for America*, (Basic Books, 1983); Harvey Brooks, "National Science Policy and Technological Innovation," in Ralph Landau and Nathan Rosenberg, eds., *The Positive Sum Strategy: Harnessing Technology for Economic Growth* (Washington: National Academy Press, 1986), pp. 154–62, and the references cited on pp. 162–67; and Kim B. Clark, W. Bruce Chew, and Takahiro Fujimoto, "Product Development in the World Auto Industry," *Brookings Papers on Economic Activity, 1987:3*, pp. 729–81.

36. Smith and Karlesky, *State of Academic Science*, vol. 1, p. 19.

37. The following discussion relies on Willis H. Shapley and others, *Research and Development in the Federal Budget: FY 1977* (Washington: American Association for the Advancement of Science, 1976), p. 16; *FY 1978*, pp. 16–17; and *FY 1979*, pp. 1–3, 21.

38. Frank Press, "Science and Technology in the White House," *Science*, January 9, 1981, p. 142.

39. Willis H. Shapley and Don I. Phillips, *Research and Development AAAS Report IV: Federal Budget FY 1980, Industry, International* (Washington: American Association for the Advancement of Science, 1979), pp. 1–16.

40. Shapley and Phillips, *Research and Development, FY 1980*, p. 14.

41. Willis H. Shapley and others, *Research and Development AAAS Report V: Federal Budget FY 1981, Industry, Universities, State and Local Governments* (Washington: American Association for the Advancement of Science, 1980), pp. 7–8.

42. Willis H. Shapley and others, *Research and Development AAAS VI, FY 1982: New Directions for R and D: Federal Budget—FY 1982, Industry, Defense* (Washington: American Association for the Advancement of Science, 1981), pp. 4–6.

43. James D. Carroll, A. Lee Frischler, and Bruce L. R. Smith, "Supply Side Management in the Reagan Administration," *Public Administration Review*, vol. 45 (November–December 1985), p. 807.

44. Shapley and others, *Research and Development, FY 1982*, pp. 7–8, 11–19. The projected civilian cuts came most heavily in the applied research programs of NASA, the Department of Energy, the Department of Commerce, and a number of the smaller civilian R&D agencies, illustrating the administration's strong commitment to market forces as the best means to commercialize research.

45. David A. Stockman, "The Social Pork Barrel," *Public Interest*, vol. 10 (Spring 1975), pp. 3–30.

46. Dickson, *New Politics of Science*, pp. 13–14.

47. Barbara Culliton, "The Private University of NIH?" *Science*, March 18,

1988, pp. 1364–65; and Culliton, "Fixing the NIH: The 110% Solution," *Science*, March 25, 1988, pp. 1479–81.

48. Jerome B. Wiesner, "The Rise and Fall of the President's Science Advisory Committee," in William T. Golden, ed., *Science and Technology Advice to the President, Congress and Judiciary*, (Pergamon, 1988), pp. 372–84. Wiesner regards the failure of the science adviser to report directly to the president as a great mistake and a virtual guarantee that the position would have little influence.

49. See S. J. Buchsbaum, "On Advising the Federal Government," in William T. Golden, ed., *Science and Technology Advice to the President, Congress, and Judiciary* (Pergamon, 1988), pp. 71–73.

50. For discussion of the arguments, see "Science Focuses on the New Presidency," *Science*, April 22, 1988, pp. 385–86. See also George A. Keyworth II, "Science Advice during the Reagan Years," and William R. Graham, "The Role of the President's Science Adviser," in Golden, ed., *Science and Technology Advice to the President, Congress, and Judiciary*, pp. 182–203, 152–54.

51. Buchsbaum, "Advising the Federal Government," pp. 71–72.

52. Shapley, Teich, and Breslow, *Research and Development, FY 1982*, pp. 11–12.

53. Shapley, Teich, and Breslow, *Research and Devlopment, FY 1982*, p. 12.

54. Shapley, Teich, and Breslow, *Research and Development, FY 1982*, p. 13.

55. A delegation of leaders and social scientists, responding to the threat posed by the Stockman-inspired assault, met at Rockefeller University in 1981 and decided to form the Consortium of Social Science Associations as a lobbying and representative arm for the social sciences under the directorship of Roberta Miller (who later became head of the NSF's social science program). COSA did much to mobilize support for the social sciences within the scientific community and to neutralize opposition to the social sciences within the Reagan administration.

56. Willis H. Shapley and others, *Research and Development: AAAS Report VII: Federal Budget, FY 1983, Impact and Challenge* (Washington, 1982), pp. 13–19; and Office of Science and Technology Policy, *Annual Science and Technology Report to the Congress: 1982* (1982).

57. Shapley and others, *Research and Development, FY 1983*, pp. 14–15.

58. *Weekly Compilation of Presidential Documents*, vol. 19 (January 31, 1983), pp. 108–10.

59. Quoted in Dickson, *New Politics of Science*, pp. 11–12.

60. *Weekly Compilation of Presidential Documents*, vol. 19 (March 28, 1983), pp. 442–48.

61. Intersociety Working Group, *AAAS Report XII: Research and Development, FY 1988* (Washington: American Association for the Advancement of Science, 1988), p. 9; and Albert H. Teich, *R and D in the 1980s: A Special Report* (Washington: American Association for the Advancement of Science, 1988), pp. 11–13, 27. See also Kenneth Flamm and Thomas L. McNaugher, "Rationalizing

Technology Investments," in John D. Steinbrunner, ed., *Restructuring American Foreign Policy* (Brookings, 1988), p. 148.

62. U.S. Office of the Under Secretary of Defense, *Bolstering Defense Industrial Competitiveness: Preserving Our Heritage, the Industrial Base, and Securing Our Future*, Report to the Secretary of Defense (Department of Defense, 1988).

63. Office of Science and Technology Policy, *Annual Science and Technology Report to the Congress, 1982* (1982), p. 11.

64. Joseph A. Pechman, *Federal Tax Policy*, 5th ed. (Brookings, 1987), pp. 134–39.

65. See statement of Norman J. Lattner, director, Federal Technology Management Policy Division, U.S. Department of Commerce, before the Subcommittee on Patents of the U.S. Senate Judiciary Committee, February 17, 1987 (mimeo); E. Jonathan Soderman and Bruce M. Winchell, "Patent Policy Changes Stimulating Commercial Applications of Federal R and D," *Research Management*, vol. 29 (May–June 1986), pp. 35–38; John M. Barry "At Last—A Reagan Technology Policy?" *Dun's Business Month*, vol. 127 (April 1986), pp. 58–59; conference report on H.R. 3773, Federal Technology Transfer Act of 1986, *BNA's Patent, Trademark and Copyright Journal*, vol. 32 (October 16, 1986); Paul A. Blanchard and F. B. McDonald, "Reviving the Spirit of Enterprise: Role of the Federal Labs," *Physics Today*, vol. 39 (January 1986), pp. 42–50; and Department of Commerce, "Final Rule on Rights to Inventions Made by Nonprofit Organizations and Small Business Firms," 37 CFR 401 (March 18, 1987), p. 8562.

66. As of fall 1988 more than sixty such R&D consortia had been registered under the act. Lists of the consortia are available from the Office of Operations, Antitrust Division, Department of Justice, and also appear in the *Federal Register*.

67. Dickson, *New Politics of Science*, p. 5.

68. Averch, *Strategic Analysis*, p. 71.

69. See Office of Science and Technology Policy, *Aeronautical Research and Technology Policy*, 2 vols. (1982).

70. David C. Mowrey and Nathan Rosenberg, "The Commercial Aircraft Industry," in Richard R. Nelson, ed., *Government and Technical Progress: A Cross-Industry Analysis* (Pergamon Press, 1982), pp. 162–232.

71. President's Commission on Industrial Competitiveness, *Global Competition: The New Reality*, vol. 1 (1985), pp. 51–60.

72. President's Commission on Global Competitiveness, *Global Competition*, vol. 2, especially pp. 16–39, 113, 173–219.

73. It issued excellent updates in 1987 and 1988. See Council on Competitiveness, *America's Competitive Crisis: Confronting the New Reality* (Washington, 1987); Council on Competitiveness, *Picking Up the Pace: The Commercial Challenge to American Innovation* (Washington, 1988); and J. David Roessner, "Evaluating Government Innovation Programs: Lessons from the U.S. Experience," *Research Policy*, vol. 18 (December 1989), pp. 343–59.

74. President's Commission on Global Competitiveness, *Global Competition*, vol. 1, p. 22.

75. See Executive Order 12591, April 10, 1987, "Facilitating Access to Science and Technology," *Federal Register*, vol. 52, pp. 13414–16.

76. See Congressional Budget Office, *Using Federal R and D to Promote Commercial Innovation* (1988), pp. 65–66.

77. See Department of Commerce, *Directory of Japanese Technical Resources—1987*, National Technical Information Service PB87-205258 (1987); and *Report to Congress: Activities of the Federal Government to Collect, Abstract, Translate and Distribute Declassified Japanese Scientific and Technical Information, 1987–1988*, National Technical Information Service PB88-194816 (1988).

78. *United States Code Service, Lawyer's Edition*, no. 9 (September 1988), p. 4383 and following.

79. For an analysis, see Christopher T. Hill, "Technology Policy for the 1990s," Twenty-second Annual Hall Memorial Lecture, Worcester Polytechnic Institute, April 26, 1988. Hill is senior specialist in science and technology policy, Congressional Research Service of the Library of Congress.

80. Secretary of Labor's Task Force, *Economic Adjustment and Worker Dislocation* (Department of Labor, 1986).

81. Congressional Budget Office, *The Benefits and Risks for Federal Funding for Sematech* (1987), p. xv.

82. Robert B. Reich, "Behold! We Have an Industrial Policy," *New York Times*, May 22, 1988, p. 29.

83. See Stephen J. Kline and Nathan Rosenberg, "An Overview of Innovation," in Ralph Landau and Nathan Rosenberg, eds., *The Positive Sum Strategy: Harnessing Technology for Economic Growth*, (Washington: National Academy Press, 1986), pp. 275–306; Harvey Brooks, "National Science Policy and Technological Innovation," in Landau and Rosenberg, eds., *Positive Sum Strategy*, pp. 119–68; and D. Teece, "Capturing Value From Technological Innovation: Interaction, Strategic Partnering, and Licensing Decisions," in Bruce Guile and Harvey Brooks, eds., *Technology and Global Industry* (Washington: National Academy Press, 1987), pp. 65–95.

84. National Science Board, *Science and Engineering Indicators, 1987* (1987), pp. 104–05, for trends in industrial R&D spending. The 1989 figures are to be found in a forthcoming NSF report; see John Markoff, "A Corporate Lag in Research Funds Is Causing Worry," *New York Times*, January 23, 1990, p. A1.

85. William J. Abernathy and Kim B. Clark, *Industrial Renaissance* (Basic Books, 1983); and Harvey Brooks, "National Science Policy and Technological Innovation," pp. 157–67.

86. Kim B. Clark, Bruce Chew, and Takahiro Fujimoto, "Product Development in the World Auto Industry," *Brookings Papers on Economic Activity*, 3:1987, pp. 729–81.

87. Martin Neil Baily and Alok Chakrabarti, *Innovation and the Productiv-*

ity Crisis (Brookings, 1988), chap. 5, discusses the problem of white collar productivity. And see, generally, U.S. Bureau of Labor Statistics, *Productivity Measures for Selected Industries and Government Services* (Department of Labor, 1988).

88. Kenneth Flamm and Thomas McNaugher conclude that "political incentives and an encrusted organizational structure [have] had unfortunate consequences for the quality of U.S. military R&D generally and more specifically for its usefulness in generating commercial returns" and that "if the nation needs a stable effort to develop its technology base, the defense budget is not the place to lodge the effort." See "Rationalizing Technology Investments," in Steinbrunner, *Restructuring American Foreign Policy,* pp. 141, 147.

89. Office of Science and Technology Policy, *Annual Science and Technology Report to the Congress, 1982,* p. 12.

90. Marshall R. Goodman and Margaret T. Wrightson, *Managing Regulatory Reform: The Reagan Strategy and Its Impact* (Praeger, 1987), pp. 34–46.

91. See *American Textile Manufacturers Institute, Inc.* v. *Donovan,* 452 U.S. 4901, decided on June 17, 1981. The "vinyl chloride" case, 838 Fed. 2d 1224, decided by the U.S. Circuit Court somewhat modified the "cotton dust" rule by holding that the courts would, in interpreting section 112 of the Clean Air Act, employ a two-stage reasoning process in which they would first consider only health data in determining the existence of a risk and then in a second stage allow benefits and costs to be weighed in determining appropriate remedies. On October 19, 1988, the Environmental Protection Agency issued a proposed rule that sought to clarify cost-benefit analysis in the context of section 408 and section 409 of the Food Additives Act (the Delaney Clauses). See *Federal Register,* October 19, 1988. A penetrating analysis of the judicial role in scientific disputes is David M. O'Brien, *What Process Is Due? Courts and Science Policy Disputes* (Russell Sage Publications, 1987).

92. Philip B. Heymann, *The Politics of Public Management* (Yale University Press, 1987); and Anne M. Burford, *Are You Tough Enough?* (McGraw Hill, 1986).

93. See William D. Ruckelshaus, "Risk, Science, and Democracy," *Issues in Science and Technology,* vol. 1 (Spring 1985), pp. 19–38.

94. Steven Weisman, "Watt Quits Post: Reagan Accepts with 'Reluctance,'" *New York Times,* October 10, 1983, p. 1.

95. Goodman and Wrightson, *Managing Regulatory Reform,* pp. 37–38.

96. Warren E. Leary, "Reagan in Switch Agrees to Plan on Acid Rain," *New York Times,* August 7, 1988, p. 1.

97. Goodman and Wrightson, *Managing Regulatory Reform,* pp. 135–36.

98. Max Friedersdorf, head of the White House legislative liaison team in the first term, remarked that "the President was determined not to clutter up the landscape with extraneous legislation." Quoted in Stephen J. Wayne, "Congressional Liaison in the Reagan White House: A Preliminary Assessment of the First Year," in Norman J. Ornstein, ed., *President and Congress: Assessing Reagan's First Year* (Washington: American Enterprise Institute, 1982), p. 50.

99. R. Shep Melnick, *Regulation and the Courts* (Brookings, 1983); and David M. O'Brien, *What Process Is Due?*

100. See I. M. Destler, "Reagan and the World: An 'Awesome Stubborness,'" in Charles O. Jones, ed., *The Reagan Legacy: Promise and Performance* (Chatham House, 1988), pp. 241–61.

101. Committee on Science, Engineering and Public Policy, *Balancing the National Interest: U.S. National Security Export Controls and Global Economic Competition* (Washington: National Academy Press, 1987), chap. 4.

102. See Strobe Talbott, *Deadly Gambits* (Knopf, 1984), for a discussion of the political dynamics underlying arms control in the first Reagan term; and Talbott, "Why START Stopped," *Foreign Affairs*, vol. 67 (Fall 1988), pp. 49–69, for an analysis of developments in the second Reagan term.

103. *Weekly Compilation of Presidential Documents*, vol. 19 (March 28, 1983), pp. 442–48.

104. Destler, "Reagan and the World," p. 250.

105. *Weekly Compilation of Presidential Documents*, vol. 19 (March 14, 1983).

106. *Congressional Quarterly Weekly Report*, vol. 45 (December 5, 1987), pp. 2992; vol. 45 (November 28, 1988), pp. 2929–33; and vol. 49 (December 5, 1988), pp. 2972–77.

107. *Special Analyses, Budget of the United States Government, Fiscal Year 1989*, pp. 1–33, J4–J5.

108. Frank Press, "The Dilemma of the Golden Age," delivered at the annual meeting of the National Academy of Sciences, Washington, April 26, 1988, p. 7. The debate was reminiscent of an earlier one that surfaced in the waning stages of the initial postwar expansionary phase of science policy. For a review of this earlier debate, see Bruce L. R. Smith, "The Concept of Scientific Choice: A Brief Review of the Literature," *American Behavioral Scientist*, vol. 9 (May 1966), pp. 27–36. Press's speech is reprinted in *Science, Technology, and Human Values*, vol. 13 (Summer–Autumn 1988), p. 224.

109. Robert M. Rosenzweig, "Thinking about Less," *Science, Technology, and Human Values*, vol. 13 (Summer–Autumn 1988), pp. 219–23.

Chapter Six

1. For a careful review see George C. Sponsler and Mark Schaefer and others, "The Case for a Department of Science and Technology," Congressional Research Service (mimeo), n.d. On November 21, 1989, Senator John Glenn, Jr., introduced a more limited variant of the proposal in S.1978, the Trade and Technology Promotion Act of 1989, which focuses on a reorganization of the Commerce Department into a Department of Industry and Technology. The bill would create within the new department an Advanced Civilian Technology Agency, modeled on DARPA in the Defense Department. An identical bill, H.R. 3833, was introduced in the House by Representative Richard Gephardt.

A more wide-ranging argument, advanced by Don E. Kash, stresses the importance of cooperation between government and the private sector in developing technology and urges the adoption of the defense model of stimulating civilian innovation through large government R&D outlays. See *Perpetual Innovation: The New World of Competition* (Basic Books, 1989).

2. Sponsler's and Schaefer's proposed department would incorporate the National Science Foundation, the National Bureau of Standards, the National Oceanic and Atmospheric Administration, the National Technical Information Service, the Department of Energy (all nonweapons energy R&D), the National Aeronautics and Space Administration, the Department of Agriculture (R&D only), and the National Institutes of Health. Total funding for the proposed department for fiscal year 1988 was to be $17.3 billion.

3. Henry Ergas, "Does Technology Policy Matter?" in Bruce R. Guile and Harvey Brooks, eds., *Technology and Global Industry: Companies and Nations in the World Economy* (Washington: National Academy Press, 1987), p. 193; and Rodney W. Nichols, "Pluralism in Science and Technology: Arguments for Organizing Federal Support for R&D," *Technology and Society,* vol. 8 (1986), pp. 33–63.

4. In absolute terms the United States continues to spend more on civilian R&D than any of its major competitors—in fact, nearly as much as Western Europe and Japan combined. See Ergas, "Does Technology Policy Matter?" pp. 191–243.

5. See Paul Kennedy, *The Rise and Fall of the Great Powers: Economic Change and Military Conflict from 1500 to 2000* (Random House, 1987).

6. J. Ronald Fox, *The Defense Management Challenge* (Harvard Business School Press, 1988), table 1.1.

7. In "Does Technology Policy Matter?" Ergas comments that policy differences "are important in shaping patterns of technological evolution, but . . . that technology policies are a *facilitating* ather than *explanatory* factor. The critical variables lie in how industry responds to the results and signals of efforts to upgrade national technological capabilities. In turn, this depends to a substantial extent on the environment in which industry operates" (p. 192).

8. Ergas, p. 198.

9. See "Reverse 'Brain Drain' Helps Asia but Robs U.S. of Scarce Talent," *Wall Street Journal,* April 18, 1989, p. 1.

10. President's Blue Ribbon Commission on Defense Management, *A Quest for Excellence* (1986), chap. 3.

11. National Research Council, *National Issues in Science and Technology* (National Academy Press, 1989), p. 86. The white paper is chapter 5 of this report.

12. Frank Press, "Annual Address to the Membership," 1989. Press cited a new optimism that the Bush administration and the nation would double the federal R&D investment in basic research from $10 billion to $20 billion in the next five years.

13. In 1984 the China Lake project was reauthorized by Congress to continue until September 30, 1990, and its 5,500-person limit was lifted. A second

project, the Federal Aviation Administration's Airway Science Curriculum program, began in 1983 in Oklahoma City. A third, approved by the Office of Personnel Management in 1985 but not implemented until 1987, was the PACERSHARE program at McClellan Air Force Base in Sacramento. A fourth was initiated by Congress directly when it passed the National Bureau of Standards (NBS) Authorization Act for Fiscal Year 1987. Other projects are pending.

14. See, for example, Linda S. Wilson, "The Capital Facilities Dilemma in the American Graduate School," in Bruce L. R. Smith, ed., *The State of Graduate Education* (Brookings, 1985), pp. 121–49; and Sean C. Rush and Sandra L. Johnson, *The Decaying American Campus: A Ticking Time Bomb* (Washington: Association of Physical Plant Administrators of Universities and Colleges, 1989).

15. The legislation includes the University and Small Business Patent Act of 1980 (and the 1984 amendments to it), the Small Business Innovation Development Act of 1982, and the Technology Transfer Act of 1986.

16. See Stephen P. Strickland, *Politics, Science, and Dread Disease: A Short History of the United States Medical Research Policy* (Harvard University Press, 1972).

17. J. L. Heilbron and Daniel J. Kevles, "Finding a Policy for Mapping and Sequencing the Human Genome: Lessons from the History of Particle Physics," *Minerva*, vol. 26 (Autumn 1988), pp. 299–314.

18. See, for example, Deborah Shapley and Rustum Roy, *Lost at the Frontier: U.S. Science and Technology Policy Adrift* (Philadelphia: ISI Press, 1985).

19. Shapley and Roy, *Lost at the Frontier*, p. 152.

20. Michael L. Dertouzos, Richard K. Lester, and Robert M. Solow, *Made in America: Regaining the Productive Edge* (MIT Press, 1989), pp. 154, 157.

21. Thirty states spent $700 million in 1986 on technology initiatives, including research parks, incubator programs, and technological training. See National Academy of Engineering, *Technological Dimensions of International Competitiveness* (National Academy Press, 1988), p. 48. For a general review, see B. Jones, *State Technological Programs in the United States* (St Paul: Governor's Office of Science and Technology, Minnesota Department of Energy and Economic Development, 1986); and David Osborne, *Economic Competitiveness: The United States Takes the Lead* (Washington: Economic Policy Institute, 1987).

22. Stephen Lock, *A Delicate Balance: Editorial Peer Review in Medicine* (Philadelphia: ISI Press, 1986).

23. National Institutes of Health, *Report of the NIH Peer Review Committee* (Bethesda, Md., 1988).

24. See Erich Bloch, "Memorandum to Members of the National Science Board," *Annual Report on the Foundation's Use of Peer Review*, NSB-89-33 (Washington, 1989); National Institutes of Health, *Report of the NIH Peer Review Committee* (Bethesda, Md., 1988); U.S. House of Representatives, *Report of a Special Subcommittee of the Interstate and Foreign Commerce Committee* (1966); *Report of the House of Representatives Select Committee on Govern-*

ment Research (1964); *Administration of Research Grants in the Public Health Service,* Committee Print, House Committee on Government Operations, 90 Cong. I sess. (GPO, 1967); *Biomedical Science and Its Administration: A Study of NIH* (1966); *Report of the Secretary on the Management of NIH Research Grants and Contracts* (1966); NIH, *Grants Peer Review: Report to the Director, NIH, Phase I* (Bethesda, Md.: Department of Health, Education, and Welfare, 1976); and NIH, *Report of the Commission on Research* (Chicago: American Medical Association, 1967).

25. NIH, *Report of the NIH Peer Review Committee,* p. 30.

26. NIH, *Report of the NIH Peer Review Committee,* pp. 30–33.

27. NSF, mimeograph accompanying *1989 Annual Peer Review Report,* pp. 11, 15.

28. General Accounting Office, *University Funding: Patterns of Distribution of Federal Research Funds to Universities* (1987), table 2.1.

29. NSF mimeograph, pp. 2, 3.

30. *Report of the NIH Peer Review Committee,* p. 27; and NIH, *NIH Fact Book, 1988* (Washington, 1988), table 17.

31. National Institutes of Health, Division of Research Grants, *Extramural Trends, FY 1979–1988* (Bethesda, Md., 1989), p. 6.

32. Benjamin N. Friedman, "Saving, Investment, and Government Deficits in the 1980s," in Bruce R. Scott and George C. Lodge, eds., *U.S. Competitiveness in the World Economy* (Harvard Business School Press, 1985), table 11-2.

33. See Edward F. Denison, *Estimates of Productivity Change by Industry: An Evaluation and an Alternative* (Brookings, 1989), p. 62; and Ergas, "Does Technology Policy Matter?"

34. Dertouzos, Lester, and Solow, *Made in America,* p. 68.

35. Quoted in William D. Ruckelshaus, "Risk, Science, and Democracy," *Issues in Science and Technology,* vol. 1 (Spring 1985), p. 22.

36. Lester B. Lave, ed., *Quantitative Risk Assessment in Regulation* (Brookings, 1983).

37. Joseph G. Morone and Edward J. Woodhouse, *Averting Catastrophe: Strategies for Regulating Risky Technologies* (University of California Press, 1986), p. 175.

38. Laurie Hayes, "Chemical Firms Press Campaign to Dispel Their 'Bad Guy' Image," *Wall Street Journal,* September 20, 1988, pp. 1, 30.

39. Science Advisory Board, *Future Risk: Research Strategies for the 1990s* (Environmental Protection Agency, 1988). See also James D. Carroll, Brian Mannix, and Stan Humphries, eds., *The Future of Biotechnology* (Brookings, forthcoming), for an example of how improved use of science has led to a regulatory framework broadly applicable to industry and the environmental movement.

40. John W. Kingdon, *Agendas, Alternatives, and Public Policies* (Little, Brown, 1984), p. 131.

41. See, for example, Lawrence D. Brown, *New Policies, New Politics: Gov-*

ernment's Response to Government's Growth (Brookings, 1983); Robert A. Katzmann, *Institutional Disability* (Brookings, 1986); Martha Derthick and Paul Quirk, *The Politics of Deregulation* (Brookings, 1985); Hugh Heclo, *Modern Social Politics in Britain and Sweden* (Yale University Press, 1974); and Pietro Nivola, *The Politics of Energy Consumption* (Brookings, 1986).

Index